FORD & FORDSON TRACTORS

Ford

FORD & FORDSON TRACTORS

Michael Williams

Special photography by
Andrew Morland

FARMING PRESS

First published in the U.K. 1985 by Blandford Press
Reprinted 1989, 1990, 1992 by Farming Press

British Library Cataloguing in Publication Data

Williams, Michael, *1935 Nov. 4-*
Ford and Fordson tractors.
1. Fordson tractors—History
I. Title
629.2'25 TL233.5

ISBN 0 85236 202 1

Published by Farming Press Books
Wharfedale Road, Ipswich IP1 4LG

Distributed in North America
by Diamond Farm Enterprises,
Box 537, Alexandria Bay, NY 13607, USA

Printed and bound in Great Britain by
Butler and Tanner Ltd, Frome and London

Contents

Acknowledgements

I would like to thank everyone who has helped me to obtain the photographs and the information which have made this book possible.

David Crippen and Cynthia Read helped me to find all of the material I required in the Ford Archives at the Henry Ford Museum and Greenfield Village, and Larry Weis and Mike Still provided all the material I requested from the Ford Motor Company in the U.S.A. and in Britain.

Mal Brinkmann of the Fordson Tractor Club in Australia and Arthur Battelle of the *Fordson and Old Tractor Magazine* in Britain also provided material and information.

Andrew Morland took many of the photographs needed for the book, with the help of Messrs. M.J. Bettles, A.G. Bray, Tony Dickinson, Graham Morley and B.R. Poole, who all made their Fordson and Ford tractors available for the camera; their help and cooperation is very gratefully acknowledged.

Andrew Morland's photographs are colour plates 3, 4, 5, 7, 8, 9, 11, 13, and 17 and the photographs on pages 64, 65, 72, 76, 77, 84, 94, 95, 97, 104, 108 and 109.

Other photographs were provided by Peter Adams, County Tractors, Crown Copyright, Ford New Holland, Henry Ford Museum and Greenfield Village, Howard Rotavator Co, Massey-Ferguson, Museum of English Rural Life, National Institute of Agricultural Engineering and David Williams. M.W.

Introduction

'To make farming what it ought to be, the most pleasant and profitable profession in the world' HENRY FORD

Tractors were an important part of Henry Ford's life. The first in a series of experimental designs was completed in 1907, to be followed later by the Fordson—the most important tractor in the history of power farming.

Henry Ford's interest in the land developed during his childhood on the family farm in Michigan. It was an interest which was never far from his thoughts, and it encouraged his determination to design a new type of tractor which would help to provide the power for a more efficient way of farming.

Enough books have already been written about Henry Ford, his company and his Model T cars to fill a small library, but these rarely allow much space to describe his interest in tractor development. This book is about the Fordson and Ford tractors which have made such an important contribution to farm mechanisation.

Tractor power has helped to raise the living standards of millions of people throughout the world, and offers the most effective measure against starvation for the millions who are still hungry. Henry Ford's tractors have helped to ensure that agriculture has the efficient power it needs.

Ford and Farming

Henry Ford's interest in tractor development was a direct result of his childhood experience of life at home on the family farm. He was familiar with a farming system which was geared to the slow pace of horses and which was often laborious. As a result he realised that the tractor offered the hope of a better way of life.

John Ford, Henry's grandfather, had emigrated from a small farm near Cork in Ireland to begin a new life in America in 1847. Some of his relations were already living near Detroit, and this was the area to which John Ford chose to take his family.

In 1848 he bought 80 acres of land near what is now the outskirts of Dearborn for $350, and established himself as a farmer. John later sold some of this land to one of his sons, William, who wanted to begin farming on his own. William Ford bought additional land, as the opportunity arose, and became a relatively prosperous member of the local farming community.

Henry Ford was born in 1863 and was William's eldest son. His father assumed that he would follow in the family tradition of farming, and Henry was expected to take an increasing share of the work as he grew older. But instead of accepting his situation as part of the labour force for the family farm, Henry Ford found the work tedious and unrewarding. He said later:

'I have walked many a weary mile behind a plow and I know all the drudgery of it. What a waste it is for a human being to spend hours and days behind a slowly moving team of horses.'

Although Henry Ford became dissatisfied with the narrow horizons of the farm, and resented the slow pace of farm work, he maintained a lifelong interest in agriculture and became a farmer and a landowner on a very large scale.

He farmed land in England, and also bought a large farm in Michigan, which was highly mechanised and was used for field testing some of his tractor ideas. Through the Ford Motor Company he also owned more than 700,000 acres of standing timber in Michigan. Then, in 1927, he bought 2.5 million acres of land in Brazil in order to secure his own supply of rubber for making tyres (tires).

Henry Ford's outstanding childhood aptitude was for anything mechanical. This interested him much more than schoolwork, in which his achievements appear to have been modest.

Charles Sorensen, who worked closely with Ford and was particularly involved in his tractor development and production, comments in his book *Forty Years with Ford* that he was astonished by Henry Ford's poor spelling:

'As a farm boy, he had no chance to go beyond the rural school, where reading and writing and arithmetic in simplest form were all a pupil could get. Even at that, he evaded much education that such a school could have offered; tinkering with watches and with machinery around the farm appealed to him more,' said Sorensen.

Henry Ford in 1906.

In December 1879 Henry Ford left home to take a job in Detroit and to serve an apprenticeship. He worked for several engineering companies, gaining experience and supplementing his wages by doing watch and clock repairs in his spare time.

He returned home for a while, married a farmer's daughter and apparently tried his hand at farming again. He also made a living by cutting and selling timber, and did some work as a service mechanic for a steam engine manufacturer before moving back to Detroit again to work for the Edison Illuminating Co.

Meanwhile, he had developed another spare time occupation. This was designing and building gasoline or petrol engines. His first success came about 1893, when he completed an engine which would run reasonably well. This encouraged him to try building a motor car.

It seems ironic that Henry Ford, who later revolutionised production methods in the motor industry, built his first car in a brick shed at the bottom of his garden. When the car was finished it was too big to fit through the doorway of the shed, and Ford had to knock down part of the wall before he could give the car its first test run in 1896.

The first Ford car, called a quadricycle, weighed only 500 lb and was powered by a two cylinder engine developing about four hp. In its original form the car had no brakes; and presumably Henry Ford placed a low priority on stopping power, as the first production version of the Fordson tractor also lacked brakes.

The motor car experiments attracted local interest and some financial backing from Detroit businessmen. Henry Ford resigned from his job at the Edison company in about 1899 in order to concentrate on car development work. After two false starts, the Ford Motor Company was established in June 1903 with ten shareholders and $100,000 of issued stock.

Within just a few years the company was well on its way to being one of the greatest success stories in commercial history. However, during the first few months of operation it was hovering near the brink of financial disaster. The speed of the Ford company's recovery is indicated by the fact that in 1908 there were no less than six dividend payments of $100,000 each, plus an additional stock issue worth $1.9 million for the shareholders.

1908 was the start of an avalanche of success for the company, and it was also the first production year for the Ford Model T car.

Fordson

THE FARMERS POWER PLANT

"To make farming what it ought to be the most pleasant and profitable profession in the world."

That is Henry Ford's vision of the Fordson Tractor and what it means to the farmer. The farmers of America have done wonderful work. They have labored hard and patiently and their efforts have made prosperity commonplace for the nation.

The limitless forces of gasoline, kerosene and electricity are now ready to loose the bonds of long hours in the field, uncertain crops and shortage of labor.

The farmer's wife can now enjoy more of the beauties of life. The tractor will make it hard to keep the boys and girls away from the farm. Conveniences now commonplace in the cities are brought to the farm and farm house by the tractor.

Mr. Ford bought thousands of acres of land, experimented for years on 62 different models of tractors at a cost of millions of dollars, before he found in the Fordson Tractor a machine he had proven a success. Mr. Ford did the experimenting with his own money. The Fordson is ready to do your work.

Ask your Fordson dealer to show you the Fordson. There are many Fordson owners near you. Ask them what their tractor has done for them.

Made By

HENRY FORD & SON, INC.

Dearborn, Mich.

Henry Ford's power farming philosophy, as it appeared in a 1925 advertisement.

Henry Ford's Model T brought a new era of cheaper motoring to millions of families, many of whom had probably never even dreamed of being able to afford a car. It was rugged and reliable, as well as being cheap, and it was especially successful in rural North

The first quadricycle, with Henry Ford at the wheel in 1904.

America where it had considerable social impact.

Between 1908 and the end of production in 1927, the company sold more than 15 million Model T cars, vans and trucks. They established the Ford company as easily the biggest motor manufacturer in the world, and helped to lay the foundations for other products including the immensely successful Trimotor civil aircraft, Liberty engines built for the U.S. Government during the First World War, and the Fordson tractor.

The commercial success was also linked to an imaginative and comprehensive programme of benefits for the company's employees. These were introduced to meet Henry Ford's ideal of giving those who built the cars a greater share in the profits they helped to produce.

In January 1914 the company made a sensational announcement of an increase in the minimum wage from about $2.60 to $5 a day. This applied to most of the company's 14,000 or so employees. There were also substantial developments in areas such as employee training, medical care and security of employment.

It was a programme which set new standards in industrial relations, and it achieved immense publicity for the company and for Henry Ford personally.

Tractor Development

Tractors were high on Henry Ford's list of priorities in the early years of his career as a motor manufacturer, and in spite of the demands made by the rapid growth of the company, he was able to find time to develop his ideas about tractor design.

He was talking about his plans for making farm tractors in 1905, according to Charles Sorensen, and he put a small team to work on a first experimental prototype in 1906 or 1907. The leader of this group was Joseph Galamb, a Hungarian by birth, who had joined the Ford company in 1905 as a trained engineer. Galamb was later described as the company's chief engineer and he was closely involved in most of the big development projects, including the Model T car and the Fordson tractor.

According to Galamb, the development work for the first experimental tractor started in a hurry with an instruction from Henry Ford that the machine should be built in three days. Sorensen's recollection suggests a more leisurely approach, with various alternative designs being considered before the tractor was built.

Galamb and his team completed the tractor in 1907, and it was tested with a wide range of equipment. Henry Ford took a close interest in the development work, and there is a surviving photograph which shows him at the controls of the tractor while it was being used on the family farm, where he had once worked with horses.

The 1907 tractor made use of a large number of Ford car components including a planetary gearbox from a Model B. The power unit, with its four water-jacketed cylinders, was also from the Model B and developed 24 hp. A basically similar engine had been used in 1904, when Henry Ford had set a new American speed record by driving a specially prepared car at 91.4 mph in order to attract publicity for his company.

Steering gear for the tractor came from a Model K car, including the high steering column, but the rear wheels had originally been made for a farm binder. The front wheels were probably from a Model K car. The tractor weighed 1,500 lb.

The details of the tractor design are less important than the general concept which it represents. Henry Ford's objective in his tractor development was the same for the cars he was building. He wanted a product which could be built in large numbers to sell as cheaply as possible. This was based on the idea that there was a vast, untapped market available if the price could be reduced sufficiently without sacrificing reliability and durability.

Design work for the 1907 tractor was carried out by a team of people, but Henry Ford provided the concept and the objectives, and he also checked the details and gave the final approval.

Henry Ford at the wheel of his first experimental tractor.

The first tractor as it appears in the Henry Ford Museum.

Using car components for the 1907 tractor may have been largely a matter of convenience, but it also fitted in with Henry Ford's idea that it would help to raise production volumes and reduce the cost if major items such as engines could be shared by both cars and tractors.

Another important feature of the 1907 design was its small size. American manufacturers were still obsessed with the idea that a tractor must be big and heavy, and the market was dominated by machines which were clumsy, expensive substitutes for a steam traction engine.

Tractor pioneers in Britain, such as Dan Albone, H.P. Saunderson and Prof John Scott, had already realised that smaller, lighter models could work effectively and offered much more versatility than the heavyweights. Henry Ford shared this view and was quite unimpressed by the fashion for big tractors.

The first experimental Ford worked reasonably well. Henry Ford called it his 'automotive plow,' and he arranged for further tractors to be built in an assortment of shapes, most of them using engines from the Model T production line.

Another of the early experimental designs, again using a transverse engine position.

During the early years of the First World War, the development programme was given added priority. This was a time when the tractor industry in the U.S.A. was growing rapidly, with a succession of manufacturers moving into the market for the first time to take a share of the expanding demand.

At this stage in the growth of the company Henry Ford had immense financial resources and plenty of talented people to develop a new tractor and put it into production. What he lacked was the support of his fellow directors, who were unwilling to venture into the farm tractor business when the company was achieving so much success with the Model T car.

Ford Motor Company was still owned by a small group of shareholders, with a board of directors to run the organisation. This arrangement had started when Ford needed outside capital to form the company, but there was no longer any financial need for the situation and Ford resented the limitations imposed on his freedom of action.

Transversely mounted, four-cylinder engine used in the 1907 tractor.

There is no date on this photograph of one of the numerous experimental tractors.

When his fellow directors failed to share his enthusiasm for the tractor project, Henry Ford decided to set up a completely separate company in order to continue the research and development programme and, eventually, to manufacture and market the tractors. The new company was named Henry Ford & Son, Inc and it began operating in 1915 from premises in Dearborn.

The new company was formally registered in July 1917 with an issue of 10,000 shares with $1 million value. The shareholders were all members of Henry Ford's immediate family, with Ford as president. In 1919, after Henry Ford had been able to buy out all the stock held by outside shareholders, Ford Motor Co. and Henry Ford & Son both came under his personal ownership. The tractor company, Henry Ford & Son, ceased its separate existence in 1920.

In 1915, Henry Ford was spending most of his time at Dearborn on the tractor project. He had transferred Charles Sorensen as his chief executive, and the engineering team was headed by Eugene Farkas who, like Joseph Galamb, was a Hungarian by birth and an engineer by training.

Farkas had a distinguished career in the Ford organisation and was prominent in some of the most ambitious development projects. He designed a 6-ton tank for the U.S. Army, was at one time in charge of the engineering team working on the Model T car replacement, and he was also the leading engineer in an abortive attempt in the early 1920s to make a commercially acceptable 8 or 12-cylinder car engine with the cylinders in an X formation.

Model T influence is evident in this experimental design.

This appears to be a rowcrop version of the design in the previous photograph, with the addition of a canopy.

By 1915 Henry Ford had become a farmer again, having bought a large area of land in the Dearborn area. This provided a convenient location for field test work, and it also meant that land was available for testing a wide range of the tractors which were then in production. According to Sorensen 'Mr Ford had acquired just about every available make and tried them out on his farm.'

This evaluation of competitive makes may have included the Wallis Cub tractor, the first production model to break away from the traditional frame design by using a one-piece curved steel structure to support and protect the main components. This was a significant advance, as it provided a rigid, dirt-proof base to carry the engine and gearbox, and helped to simplify the production process.

The Farkas design which emerged from the 1915 development programme also had a frameless construction, but on the Ford tractor this was achieved in a completely different and more simple way. The Ford engine, transmission and differential housing were joined together to form a rigid unit which performed the same function as a frame, and also enclosed the complete mechanism to keep out dirt and water. All three of the main units were housed in cast iron, and were designed for production line assembly techniques.

The Henry Ford & Son badge on the Farkas design in 1916.

A 1916 photograph of the Farkas design.

HENRY FORD
&
SON

While Eugene Farkas and his team were working on their frameless design, Charles Sorensen was bringing in more equipment so that the tractor company could expand its operations as a completely self-contained unit. Sorensen approved of the design the engineering team had developed, and he helped to convince Henry Ford that they should take the project through to the field test stage.

Additional workers were hired, and a batch of about 50 tractors was built for evaluation. Some of these were taken to the Ford farms, where numerous photographs were taken. The photographs which have survived show that the pre-production prototypes closely resembled the Fordson tractor which arrived in 1917.

Some of the photographs taken in 1916 show the name, Henry Ford and Son, in a circular pattern on a plate at the front of the tractor. This was later simplified to Fordson for the production model.

The name Ford, which was used for the Model T car and was already one of the best known trade names in the world, was not at this stage on the tractors because they were not a Ford Motor Company product. There is also a theory that Henry Ford was prevented from using the name Ford because it was already used as the trade name for a tractor built by the Ford Tractor Company of Minneapolis.

A harvesting scene in 1917, with three of the Farkas tractors.

This view of tractor 28, from the batch of 50 built to the Farkas design, shows that the steering wheel was offset.

The Minneapolis Ford was a three-wheel design which was on the market from 1915 to 1918. The tractor was backed by a group of businessmen who, it is said, hired someone named Paul Ford so that they could use his surname for their new tractor.

The idea, according to one theory, was to associate the tractor from Minneapolis with the company which was building the immensely successful Ford cars. Certainly the name was used with considerable prominence, and may have helped to influence a few additional sales.

It is much less likely that the Ford name on another company's tractor would have influenced Henry Ford's choice of name for the tractors he was producing. It is also unlikely that the Minneapolis company was strong enough even to attempt to influence Henry Ford; that company was close to financial collapse when the Fordson tractor was ready for production.

Although the Henry Ford and Son tractors performed well in their field test programme, there was no attempt to rush them into production. They attracted plenty of publicity, and one tractor took part in a big ploughing demonstration at Fremont, Nebraska in 1916. Henry Ford was quoted in the press as having promised to build 10 million tractors, and there were also reports that the new tractors would sell for $200.

Meanwhile, thousands of farmers who wanted cheap tractor power were able to use their Model T Ford cars. Car conversion kits sold in large numbers, particularly in the U.S.A., with sales reaching a peak between 1916 and 1920. C.H. Wendel lists 45 kit manufacturers in his book *Encyclopedia of American Farm Tractors*. This covers the manufacturers who were active in 1919, and omits some which had appeared briefly in previous years, as well as some British kit companies. Most of the kits were designed for the Model T, as this was the most popular type, especially in rural areas where it was a firm favourite with farming people.

According to the advertisements, it was a quick and easy job to use a kit to convert a car into a tractor unit in order to plough a field—and equally simple to change back again so that the car could carry the farmer and his family to church on Sunday. It is difficult to believe that it was really quite *so* straightforward and convenient. Indeed, it seems likely that the kits were often used as a more permanent conversion for cars which were past their prime and were able to finish their days on the plough.

A Model T tractor conversion pictured in Britain about 1917.

An Eros conversion on a Model T pictured recently in Canada.

Most of the kits included large diameter wheels with lugs for traction, and a mounting frame designed to keep the car level when the big wheels were in use. There was usually some mechanism to reduce the forward speed for field work, and a modification to increase the cooling capacity of the engine to prevent overheating.

Obviously, this form of tractor power had distinct limitations, and it is significant that the kits vanished from the market as soon as tractor prices began to fall substantially in the early 1920s. It is also obvious that the kits were meeting a genuine need among farmers in the U.S.A. and in Britain, and it is a tribute to the ruggedness of the Model T Ford that it would stand up reasonably well to such a demanding job as ploughing.

One of the leading American firms in the conversion kit market was E.G. Staude Mfg Co of St Paul, Minnesota. They made the Eros conversion which was popular in the United States and, for a while, in Britain. It was claimed that the Eros conversion gave a six-fold increase to the Ford engine cooling capacity. This was achieved by using an impeller to speed up the circulation of cooling water, and by providing a more efficient radiator with extra water capacity. There was also a force-feed lubrication system for the Model T engine. Conversion kits of this type cost from about $180 to $250.

One of the many people who saw a future for the Eros conversion was Harry Ferguson. At the time he was running a garage business in what is now Northern Ireland, with the agency for the Waterloo Boy tractor.

The Trafford Model T conversion was designed in Britain.

Ferguson had become familiar with many different makes and models while touring farms at the request of the Irish Board of Agriculture to help improve tractor utilisation. In 1917, he started work on a new method for attaching implements to tractors. This development work eventually produced the Ferguson System, but it started with a plough designed for a Model T car with an Eros conversion. The new plough was demonstrated in February 1918 in a 12-acre field near Coleraine in County Londonderry.

The most successful British made kit was sold by the Trafford Engineering Co of London, and was based on the design of James Hodgson of West Walls, Carlisle. His kit included a triangular subframe, with the apex of the triangle attached to the middle of the Model T front axle. The base of the triangle carried an axle with three wheels, but there was a fourth wheel in the kit which could be added to help achieve extra grip in difficult conditions.

The manufacturers claimed that it took just 30 minutes to convert the car for tractor work, and another 30 minutes to convert back again. They further claimed that the Trafford, also referred to as the Tracford, would pull a two-furrow plough at 2 mph, and was capable of ploughing three acres a day.

A Trafford tractor unit was entered in the Scottish tractor trials of 1918, and it made a favourable impression on the judges. 'The whole combination gives a very excellent adhesion. The attachment appears to be well designed and is certainly effective,' said the official report.

A Model T Ford with an Eros conversion shown with the plough which was the first development stage for the Ferguson System.

Henry Ford may have welcomed the mushrooming market for conversion kits as another endorsement for the sturdy reliability of his Model T and he also, no doubt, welcomed the extra publicity which the tractor conversions provided. But his own tractor developments continued to concentrate on the Farkas design of 1915–16, which would eventually help to sweep all the conversion kits off the market.

Fordson

The Fordson in Production

While the tractor development programme continued in the U.S.A., the First World War in Europe was dragging into its third year and Britain was facing a food crisis.

When the War started, there had been a widespread assumption that the fighting would end in a swift Allied victory. This early optimism was gradually replaced by the grim reality of slaughter measured in years rather than months.

By the middle of 1916 it had become obvious that food supplies were a major factor in the war, with a real possibility that Britain might be starved into submission.

For years Britain had relied on large quantities of imported food which could be shipped from countries such as Australia, Canada and the United States at prices British producers were unable to match. One result of this cheap food policy had been a switch from arable farming in Britain to livestock production, and there had also been a lack of capital for investment in agricultural equipment, including tractors and machinery.

Another result of the policy was that the ships which brought food to Britain had become an obvious lifeline which the Imperial German Navy was able to attack during the War. The U-boat campaign destroyed so many merchant ships that the supply of food from overseas was seriously threatened. Consequently, the British Government was forced to allocate resources to raise the level of output from their farms.

The plan involved the ploughing up of a large acreage of grassland in order that more grain could be produced to compensate for the lost imports. This meant additional work at a time when manpower in agriculture had been substantially reduced because of the War. It was estimated that more than 400,000 workers had left the land to join the British Army by the end of the War. With them went large numbers of farm horses, required by the military authorities for transport work in the battlefields of Europe.

Tractor power was essential if the ploughing targets were to be met, but the number of tractors already available on farms was small and there was no easy way to obtain the additional tractor power required.

A Reading University research project in 1984 indicates that there were fewer than 500 tractors on British farms in 1914, when the War began. Sales had been so small during the pre-War period that most manufacturers had concentrated on exporting and companies such as Clayton and Shuttleworth, Daimler, Fowler, Marshall and Ransomes had developed tractors to suit overseas conditions. These were too big and expensive to suit the needs of the ploughing campaign in Britain, and they were unsuitable for the large-scale production which was necessary.

A July 1917 photograph of a pre-production version of the Fordson.

There were also some smaller tractors which were more suited to British conditions, but they were made by companies which were too small to cope with the numbers needed by the Government. Britain's biggest tractor manufacturer at that time was Saunderson of Elstow, Bedford. In 1917, the Board of Agriculture placed an order for 400 Saunderson Universal tractors to be supplied with 400 ploughs. This was as much as the company could cope with, and a substantial amount of the work had to be sub-contracted out to other firms in order to meet the delivery dates.

Tractors were available from the United States and substantial numbers had already been imported. Some of these performed well under British conditions, including the International Harvester Titan and the Waterloo Boy and both were popular with a good reputation for reliability. There were also some which were poorly designed, unreliable and with an inadequate spares back-up.

One of the people closely involved in the food supply problem was Percival Perry, later to become Lord Perry. He was the chief executive of the Ford subsidiary company in Britain, and early in 1917 he was helping temporarily in the Food Production Department of the Board of Agriculture and Fisheries. He was aware of the urgent need for tractor power, and he also knew that the Ford tractor development project was at an advanced stage.

It was Perry who arranged for two of the prototype tractors to be shipped to Britain in January for evaluation. He was also able to organise an official request from his Department to the Royal Agricultural Society to have the test programme organised.

The tests started on April 24th and lasted two days. They were held on the Cheshire estate belonging to Sir Gilbert Greenall, with a panel of five judges appointed by the Society. The judges were Prof W. Dalby of the City and Guilds Engineering Institute, Mr F. Courtney, the Society's engineering consultant, Mr R. Greaves, chairman of the Society's machinery committee, and Mr R. Hobbs and Mr Henry Overman who were both 'Practical Agriculturists'.

The ploughing tests were carried out on heavy, weed-infested soil, and the judges considered that the quality of work achieved by the tractors could have been improved if different ploughs had been available. The work was done with a 2-furrow Oliver plough and a 3-furrow Cockshutt, and both were said to be unsuitable for the conditions in the fields provided for the tests.

When the field work was finished, one of the tractors was taken

A consignment of Universal tractors leaving Saunderson's factory near Bedford in 1917 to meet a British Government order.

to the Ford assembly plant in Manchester so that the judges could see it being dismantled for a more detailed examination.

The judges produced their official report for the Food Production Department, and it showed that the tractors had created a favourable impression. There was praise for the way the engines started easily from cold, for the small turning circle required, and for the general ease of operation. There was also enthusiastic approval for the light weight of the tractors, as this would help to reduce soil compaction, the judges said.

They described the general design as being 'of ample strength, and the work of first-rate quality'. The wheels were considered inadequate, but the report noted that an improved design had already been produced by Ford engineers in the United States. Judges warned that wheelbands would be essential if the tractors were to travel on roads while lugs were fitted to the rear wheels.

'Bearing the above points in mind we recommend, under existing circumstances, that steps be taken to construct immediately as many of these tractors as possible', the report concluded.

The RASE report and recommendation confirmed the view already held by the British Government that the Fordson tractor would be suitable for large scale production to work in the ploughing campaign. A licence had already been granted for a new Ford factory to be built at Cork in Ireland, and this was confirmed in a Parliamentary answer on March 7th. The factory was to produce 'agricultural machinery'.

By April 7th, Percival Perry had been able to send a cable to Edsel Ford requesting the loan of drawings and personnel so that the tractor could be built in Britain. The cable included a request that Charles Sorensen should be sent to Britain to help set up and run the factory, which would operate on a non-profit basis under Government control.

The reply from Henry Ford promised complete co-operation, and it was followed by prompt action to get the project underway as soon as possible. Sorensen arrived in London in May ready to help to get production started in the Cork factory, which would operate under the Ministry of Munitions.

It was Henry Ford's idea to have a factory at Cork, where he had family associations, and his objective was to make a practical contribution to help reduce the high level of unemployment in the area. The whole of Ireland was at that time under British control, and the Government in London had encouraged the factory as a way to increase industrial development in Cork.

All seemed set for the tractor project to go ahead, with Charles Sorensen supervising the arrangements. Percival Perry resigned from his Food Production Department post so that he could give more time to the tractor factory. Henry Ford was providing all the support he had promised, and he had also given an assurance that he would not make a profit out of the tractors to be built for the British Government or charge for the use of the drawings and patents.

This situation changed abruptly in June, when German aircraft

Lord Northcliffe and members of his party at the Fordson demonstration in 1917.

Lord Northcliffe at the wheel, followed by Henry Ford.

made successful bombing raids on London and other British cities. Britain needed extra warplanes to provide a defence against these attacks, and the Government made immediate arrangements for an emergency production programme. In order to provide the resources needed for the new aircraft, other production schedules had to be abandoned or delayed, and the Fordson tractor project was one of these.

Instead of using the Cork factory for the tractors, the Government asked Henry Ford if they could be manufactured in the U.S.A., to be shipped to Britain against a firm order from the Government. Henry Ford agreed to the change of plan, with a commitment to send the first 6,000 tractors to Britain.

Sorensen hurried back to supervise arrangements for the tractors to be built in the existing Ford factory in Dearborn, Michigan. At this stage there are some indications that Henry Ford would like to have delayed production because he wanted to make more changes to the design. In October, Lord Northcliffe visited Dearborn to see how the tractor project was progressing and to use his influence, as a senior representative of the British Government, to encourage Henry Ford to speed up the programme.

During the visit there was a demonstration of the pre-production version of the tractor, providing Lord Northcliffe with an opportunity to take a turn at the wheel for a test drive. The tractors appear to have made a considerable impression on his Lordship, and he later referred to the occasion and his own brief experience as a tractor driver with great enthusiasm.

He claimed that he was able to plough at between three and four miles per hour, and the tractor seemed to 'romp across country' when pulling a set of disc harrows.

Another visitor to the Dearborn factory during the pre-production period was J. Edward Schipper, a well known authority on the American motor industry. His description of the Fordson in the journal *The Automobile and Automotive Industries* was published in August and contained a wealth of technical detail which was provided by the company under direct authorisation from Charles Sorensen.

'The engine block is cast from semi-steel; in other words, from iron refined by the addition of about 15 per cent of steel scrap. The crankcase is also of semi-steel, as are the pistons, which latter are $4\frac{7}{8}$ in long. The pressure on the piston head is transmitted through the pin bearings, which are in the bosses of the piston, to a $1\frac{3}{8}$ in piston pin of chrome vanadium steel. The bearing length on this pin is $2\frac{1}{4}$ in, the pins bearing directly on the cast iron of the pistons.'

'Three crankshaft bearings are used, and are provided with interchangeable caps. This is one of the features of the machine tending towards simplicity and making for low manufacturing cost. It is also a factor in the maintenance of the tractor. All of the crankshaft bearings are of 2 in diameter and $3\frac{1}{4}$ in long'.

Schipper appears to have been impressed particularly by the quality of the materials used to make the tractor, and he noted that

Henry Ford, left, with his son, Edsel, and what the original caption claims is the first production Fordson.

Fordsons in the factory at Brady Street and Michigan Avenue in 1918.

this was an important factor in reducing the weight, which at this stage was 2,500 lb. He also mentioned that special heat treatment processes had been developed at the factory to help ensure that every component had the most suitable physical properties for its function in the tractor.

According to the Schipper report, the factory was almost ready to begin full-scale commercial production, but this proved to be optimistic. The first tractor was not completed until October 8th and only 254 had been built by the end of the year. The project was behind schedule, but the problems were quickly dealt with and output increased during 1918 to total 34,167 for the year.

The first tractors were probably too late for the 1917 autumn cultivations, but some were ready for work as soon as the ground was fit at the end of the winter.

'The long-delayed Ford tractors have arrived and each is now ploughing up England at about half an acre an hour', the *Daily Mail* announced optimistically on March 16th. 'There are hundreds now at work, and soon there should be thousands.'

In the *Daily Mail* reports the Fordsons were good news for Britain, and a correspondent who had toured the Ford factories in February described the tractor as a 'wonderful instrument of war'. Press comment was not always as enthusiastic as this, and there were some reservations about the Government deal which was bringing the tractors to Britain.

Much of the criticism came from journals closely linked to the interests of the British motor and farm machinery industries. They looked beyond the immediate food production problems and were concerned that the preference for the imported Fordson could be a serious set-back for British manufacturers.

One example of this view is the detailed analysis given in *The Motor* on November 13th, 1917, under the heading, 'The Truth About the Ford Tractor'. This stated firmly that the Government had chosen the wrong tractor. Although the Fordson was 'very well thought out from a mechanical and manufacturing point of view', and was ideal for small farms, it was not suitable for a ploughing campaign which demanded high output from an organisation with few experienced drivers.

'It is true that Mr Ford has given us the benefits of his patents—such as they are', said *The Motor*, 'and every assistance in production without charge, but he will be extremely well repaid by the enormous advertisement value which the fact of the British Government's employment of his machine will furnish, and he will get a start on his new line of manufacture which any British manufacturer would have given his ears, so to speak, to secure. In that, he and his men have been far-seeing, clever, and successful.'

American tractor manufacturers were also affected by the British deal with Henry Ford. Some, including International Harvester and the Waterloo Gasoline Engine Co., had built up a substantial export market in Britain. After the Ford agreement, all the cargo space which could be spared for tractors was needed for the Government's Fordsons, and there were new regulations to make it more difficult for other companies to ship tractors to

Fordson number 1857, one of the tractors supplied to Britain in 1918, photographed in the Science Museum, London.

Britain. For example, all companies except Ford had to produce evidence of a firm order from a British farmer for every tractor they wished to export to Britain.

These regulations became an important issue in the row which developed when the Ford Motor Co. announced that they had new Fordson tractors for sale in Britain. The tractors were priced at £250, of which £80 represented the cost of freight from Dearborn, and they were available for immediate delivery.

The tractors were part of the 6,000 ordered by the British Government for the ploughing campaign. The original intention had been to retain all of these tractors under Government ownership and control, but the 'for sale' announcement on June 11th, 1918, was a result of a policy change by the Government which brought an angry response in some sections of the British press.

An official explanation for the Government policy switch was demanded in *The Agricultural Gazette's* July issue. Farmers should have the right to choose the tractor they considered best suited to their needs, but the Government had 'queered the pitch' of all other manufacturers and importers, so that only the Fordson was available.

In the same month *The Motor* pointed out that as the Fordsons were being offered for sale, orders for them from farmers could not have been in existence when the tractors were originally imported.

Henry Ford and Charles Sorensen, both standing below the rear wheel, with one of the first of the tractors for Britain.

This must mean that the tractors were outside the official import regulations, *The Motor* claimed.

Various explanations have been offered for the decision by the British Government to allow the company to sell some of the tractors imported specifically for the ploughing campaign. One suggestion is that the hiring out scheme had proved to be a failure, and it was thought that some of the tractors could be used more effectively under private ownership. Another suggestion was that the scheme had been so successful that the Government had decided not to continue it because they lacked the manpower to cope with the expected harvest.

Meanwhile production in the Dearborn factory off Brady Street and Michigan Avenue was expanding rapidly. The British Government order was completed in April, and this made it possible to begin supplying tractors against the huge backlog of orders from Canada and the United States. It also became possible for Henry Ford to supply tractors as gifts to his close friends and to people he admired. Luther Burbank and Thomas Edison were the first people to be given new tractors, and a small batch of tractors was later presented to the royal family in Belgium to be used to help in the reconstruction of Belgian agriculture after the War.

The wartime plan to build tractors at the Ford factory in Ireland was revived, and the first Irish Fordson was completed at the Cork

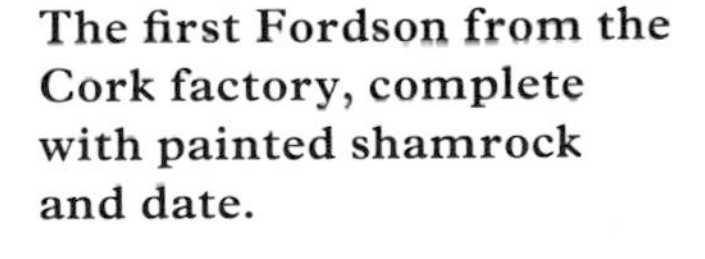
The first Fordson from the Cork factory, complete with painted shamrock and date.

More Irish celebrations as the Mayor of Cork takes the wheel.

factory on July 4th, 1919. This tractor was later presented to David Lloyd George, the British Prime Minister whose Government had authorised the establishment of the Ford factory at Cork. It was used on his farm in Wales.

Production at Cork peaked at 3626 during 1920, and fell sharply with the decline in European tractor sales during 1921–22. The Cork site had been chosen mainly for social and political reasons, in spite of the lack of raw materials and the small demand for tractors in Ireland. This meant that the tractor plant had to rely on imported components and raw materials, and almost the whole of the output had to be shipped overseas.

Cork was planned as an assembly and production plant to supply tractors to Europe, with Britain as the main market. When it became evident that tractor production alone would not provide sufficient volume to meet the factory development plans, a substantial amount of Model T car component production was added to the work load.

More serious problems for the Cork factory were developing during 1922 when the British Government was imposing trade restrictions to limit the importation of some industrial products from the recently created Irish Free State. This meant that Fordson tractors for the British market had to be sourced from the

U.S.A., which closed down the tractor operation in Ireland, until Cork returned to prominence as a tractor centre in 1929.

Henry Ford had never regarded the Dearborn factory as a long term base for his tractor production, because the vast Rouge plant was planned as the future centre for both car and tractor manufacturing. The Rouge factory was built on land Ford had bought for the purpose, with frontage on the Rouge River and with excellent rail links to every part of the U.S.A.

The tractor plant was transferred from the old Dearborn factory to the Rouge in the winter of 1920–21. A large area of the 'B' building was allocated for the tractor operation, which was still under Sorensen's direct control. The 'B' building had recently been converted from its wartime use as the factory producing Ford's Eagle anti-submarine boats for the United States Navy.

This was an important stage in the development of the Ford tractor business, because the new factory provided the space and the facilities to develop the tractor on the massive scale Henry Ford had envisaged. It meant that more of the components could be built in the tractor plant instead of being sourced from outside suppliers, and it also gave the production scale and efficiency which allowed Ford to cut his prices again and again.

Fordson production at the Rouge factory dominated the tractor market throughout the world during much of the 1920s. Production started on February 23rd, 1921; it ended there seven years later when the whole of the tractor manufacturing operation was transferred to the Cork factory. During this period, the Rouge had produced more than 500,000 Model F tractors, which were shipped to almost every part of the world, making a massive contribution to agricultural development.

Fordsons leaving Cork in the hold of a ship.

Model F in Action

Tractors shipped to Britain for the Government's food production campaign were distributed throughout the country, and were operated under the control of the County War Agricultural Committees. The Government met the costs of the fuel and spare parts, paid the wages of the drivers, fitters and local supervisors, and provided vans for fuel and parts deliveries to the farms where the tractors were at work.

A survey of the operation of the scheme in Kent was published in the *Journal* of the Royal Agricultural Society of England for 1918. This provides a detailed record of the performance of the three tractor makes principally used in the county.

The Kent War Agricultural Committee tried out 15 different makes and models during the early stages of the campaign. Most were American, including Emerson-Brantingham, Cleveland, Case, Sandusky, Samson and Whiting-Bull. Some of these were found to be unsuitable for the local conditions, or were rejected because of problems over the supply of spare parts.

During the autumn of 1918 the Committee had reduced its tractor fleet to just three different models, and was operating 112 International Harvester Titans, 64 Fordsons and 4 Overtimes. The name, Overtime, was used in Britain for the Waterloo Boy Model N, the American built tractor which was taken over by Deere & Co when they decided to move into the tractor market.

Performance figures for the Kent tractors show that the Fordsons achieved better work rates than the Overtimes and Titans, and cost less to operate. There was also less time lost for repairs for the Fordsons when ploughing, but this was not the case when working with a binder.

The figures in Table 1 are based on records kept in Kent while ploughing during the summer and autumn of 1918.

Table 1 Ploughing Performance

Tractor type	*Hours per acre ploughed*	*Costs per acre ploughed* Labour	Fuel	Grease	Oil	*Hours for repairs*
Fordson	3.81	31p	50p	0.03p	11.6p	2.23
Overtime	4.03	40p	75p	0.14p	12.2p	6.76
Titan	4.17	46p	70p	0.07p	14.8p	2.81

A publicity picture of an early 1920s Fordson.

1 The first experimental Ford tractor of 1907.

2 Fordson number 1 photographed in the Henry Ford Museum.

3 A 1918 Fordson at the Bicton Gardens Museum, Devon, England.

Opposite
5 Model N Fordson with water washer air cleaner.

4 1925 Model F.

FORDSON

7 One of the green Fordsons from Dagenham, built in 1942.

Opposite
6 Orange Fordson built in 1939.

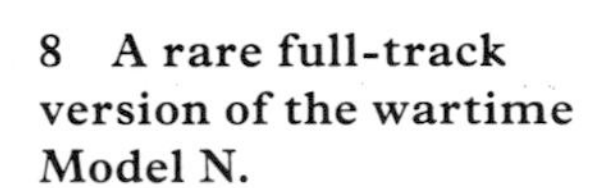
8 A rare full-track version of the wartime Model N.

9 A 1943 Model N with rowcrop wheels.

10 The front cover of the E27N Fordson sales leaflet.

Opposite
12 The cover of a 1947 sales leaflet for the Ford 8N.

11 A 1948 Fordson E27N.

13 'Ford Tractor with Ferguson System'.

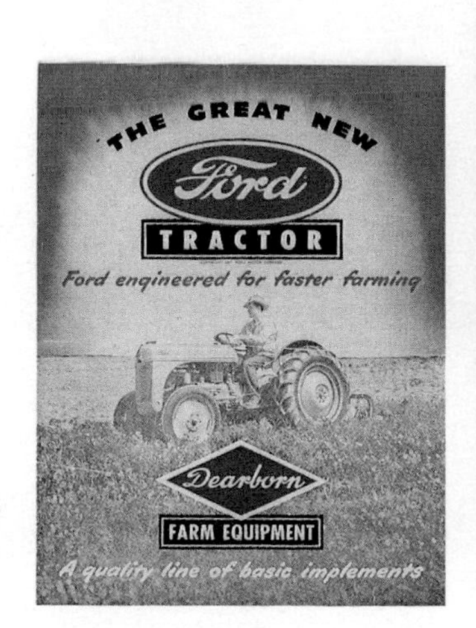

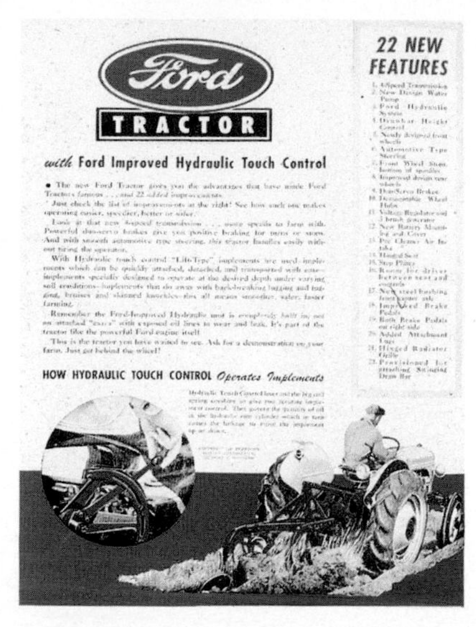

14, 15, Sales features of the Ford 8N tractor shown in the 1947 leaflet.

16 A Ford 8N tractor in the Henry Ford Museum.

An International Harvester 10–20 Titan, familiar in Britain during the war-time food production programme.

In addition to their work for the ploughing campaign, the Government tractors were also used for other work, including harvesting with binders. The Kent War Agricultural Committee kept detailed records during the 1918 harvest, and data from the report is included in Table 2.

Table 2 Harvesting Performance

Tractor type	*Hours per acre harvested*	*Paraffin*	*Oil*	*Repairs hours per acre*
		gals per acre		
Fordson	1.42	1.76	0.22	0.47
Overtime	1.76	2.12	0.20	0.16
Titan	1.70	2.09	0.29	0.29

The performance figures provide an interesting comparison between the three different tractors, and they also indicate the poor results achieved in the Government campaign. The Kent figures are fairly typical of the national situation, which attracted criticism because of the high cost of the operation. The fact that almost 200 tractors operated by the Kent Committee averaged about four hours each to plough an acre requires some explanation.

One of the biggest problems in Kent, and elsewhere in the country, was the shortage of experienced drivers. Many of the men and women employed in the scheme had little or no previous knowledge of farming or of machinery, and there was no provision in the scheme for properly organised training courses.

The learner drivers were not sufficiently conscientious about routine servicing, according to the author of the RASE report, Mr G. H. Garrad.

'Grease caps are fitted by most manufacturers,' he wrote. 'If the driver will only see that these greasers are properly filled and screwed down once or twice a day, it will increase the life of the tractor considerably. Over-greasing is a good fault, and one easily remedied.'

When replacement parts were needed the wartime situation and, perhaps, the bureaucracy of the campaign organisation caused long delays. It was not unusual, the report said, for tractors to be standing idle for several weeks because of a parts shortage.

Other difficulties were linked with the nature of the ploughing campaign, which took each tractor to a list of different farms during the season. Fuel used on the road was set against the acreage worked, producing a distortion in the consumption and cost figures. The time required for travelling from farm to farm reduced the amount of work the tractors were able to complete each week.

The British Army lends a hand with the ploughing campaign using International Harvester 8–16 Mogul tractors.

There was also a supervision problem. Some drivers, the report suggests, took advantage of lack of effective control and made less effort than would be required in a normal farm situation. Another factor was the quality of land available for the Government tractors to plough. Farmers who used the Government tractors kept the easy working land for their own ploughmen to deal with, and left the worst fields for the tractor teams.

In spite of the problems some tractors achieved more creditable performance figures, including a Fordson operated by the Kent Committee which earned the following report in the *Implement and Machinery Review* for February 1919.

'Though not of Amazonian physique, Miss Worthington, the lady tractor driver, of Romney Marsh, Kent, is endowed with an abundance of pluck and grit, and has actually established a county record in tractor ploughing.' Miss Worthington and her tractor ploughed 16½ acres in 45 hours for an average 3.1 gallons an acre fuel consumption. 'Needless to say, there is today less and less antipathy to lady tractor drivers', the report added.

Henry Ford and tractor No. 500,000 in 1925.

In order to encourage drivers to achieve better results in the ploughing campaign, an award scheme was organised in 1918 with cash prizes for the best fuel efficiency figures and for the biggest acreage ploughed. The competition was open to all drivers taking part in the campaign in England and Wales, and was based on records kept during a three-month period by local supervisors.

The fuel efficiency prize was won by a Lincolnshire driver who averaged 2.13 gallons per acre ploughed with a Fordson. Another Fordson was runner-up in this section with 2.26 gallons an acre. The prize for the biggest acreage ploughed was won by the driver of a Titan tractor.

One of the limitations of the Fordson design was lack of weight to help achieve good wheelgrip for jobs such as ploughing. This sometimes showed up when drawbar pull figures for various tractors were compared during performance tests.

British farmers were offered an opportunity to see a large number of tractors working competitively at the trials organised in Lincolnshire in 1919 by the Society of Motor Manufacturers and Traders. The event attracted 56 entries, of which four failed to arrive, and included ploughing tests on both heavy and light soils and a haulage trial which took the tractors on to public roads.

Ford was represented by a Model T with an Eros conversion, entered by the Eros distributor in Britain, and by three Model F

Tractors awaiting despatch from the Rouge in 1926.

A 1927 Industrial Fordson.

tractors. The Fordsons took part in all the field tests, but were refused a place in the haulage section because the organisers said the lack of brakes and springs made them unsuitable for road work.

On the light land the two Fordsons competing in this section of the trials achieved a fairly creditable best performance of 0.78 acres ploughed in an hour. An International Harvester Junior tractor, with a 22.5 hp engine rating, achieved 0.64 acres an hour, and a 25 hp Austin, the Fordson's main British rival, worked at 0.79 acres an hour. The Eros conversion, listed as producing 16.9 hp output, ploughed 0.49 acres an hour.

The heavy clay soil provided for part of the ploughing trials, made effective wheelgrip more difficult to achieve and this is reflected in the performance figures achieved by the two Fordsons taking part. They managed a best workrate of 0.43 acres an hour, which was the same as the result of the International Junior. The best output by an Austin in this section was 0.66 acres an hour, and the Eros did surprisingly well with 0.53.

Further evidence of wheelslip problems came when the Fordson tractor went through the test programme at the University of Nebraska in 1920. More than 60 tractors were tested at Nebraska during 1920, and only three of these weighed less than the 2,710 lb recorded for the Fordson. In the belt tests the Fordson performed well, producing more than the rated 18 hp, and achieving 6.38 hp-hours per gallon of fuel used, which was substantially better than the average for the year's tests.

In the test section to measure drawbar performance, the Fordson achieved the doubtful distinction of the highest wheelslip figure for the year at 23.8 per cent. In the ten-hour rated load test the drawbar pull was recorded as 886 lb, and the fuel consumption figure fell to only 2.45 hp-hours per gallon, the lowest fuel efficiency figure in the 1920 test series.

Later versions of the Fordson were able to deliver more power at the drawbar when spade lugs were used instead of the original angled bars to help the rear wheels grip the soil.

Another result of the Fordson's light weight was a tendency for the front end of the tractor to lift when trailed equipment, such as a plough, came up against a solid obstruction. This could happen in fields with buried rocks or tree roots, and it sometimes meant that the tractor would tip over backwards. Reynold M. Wik in his book *Henry Ford and Grass-Roots America* quotes contemporary press reports suggesting that as many as 136 tractor drivers had been killed in Fordson accidents in the U.S.A. by 1922.

The overturning risk brought several safety cut-out devices on to the market on both sides of the Atlantic. These were designed to cut the current to the spark plugs or disengage the clutch as soon as the front of the tractor began to lift. The problem of overturning was aggravated in the early days of the Fordson by the high percentage of first-time drivers using the tractors.

In spite of the disappointing drawbar pull figures and the lack of front end stability, the Fordson success story developed rapidly as the sales figures increased and the tractor began to dominate the tractor industry throughout the world. Henry Ford's aim had been to develop a tractor which could be produced so cheaply that it would bring the benefits of power farming within reach of thousands of smaller farms for the first time, and this was the Fordson's greatest achievement.

In Britain the first quoted selling price for a Fordson was £250. This was in 1918, and was for some of the tractors which had been ordered by the Ministry of Munitions. The price was raised in 1919 to £280, when the advertised price for an Austin tractor was £300. In 1920 the price of the Austin had been raised to £360, but

Rear wing modification on a 1927 Fordson.

the Fordson price had fallen to £260. Further price reductions followed, and in 1931 a British farmer could buy a new Fordson for £156, or for a first payment of £50, with a year or so to pay off the balance.

A similar price-cutting policy was operated in the U.S.A. When the British Government order had been completed, tractors were offered to American farmers for $750. This price was repeatedly reduced, to reach $395 in 1922.

The price reductions were firmly in line with Henry Ford's policy of making the tractor available to as wide a market as possible, but they also made life extremely difficult for other tractor companies in the U.S.A. and in Europe. During the 1920s, hundreds of tractor manufacturers pulled out of the market, forced out by the general depression in the farming industry and by the commercial success of the Fordson.

The Fordson success story was on an international scale, and it even penetrated the political isolation of the post-revolutionary communist regime in the USSR. In the early 1920s, the Soviet Government was trying to cope with a desperate food shortage following the civil war and a series of poor harvests. Like the British Government a few years previously, they needed tractor power in a hurry and they turned to the Fordson.

Negotiations between the Amtorg purchasing agency in the USSR and the Ford company resulted in a series of massive orders for a total of more than 26,000 Model F tractors. One batch of 10,000 was built at the Rouge factory during a seven week period at the end of 1926 'in addition to the normal production quota for domestic absorption', the company claimed.

The Russian tractor deals included large quantities of spare parts and the services of Ford personnel in the USSR to help with the organisation of parts and maintenance back-up for the tractors on the farms. The company also provided advice on ways to raise efficiency and productivity at the Krasny Putilowitz factory where the Russians were building their own copy of the Fordson tractor.

Ford managers, accustomed to American standards on the farm and in factories, were sometimes shocked by what they found in the USSR. A delegation which toured farms and service facilities at the request of the Russian authorities in 1926 found that knowledge of servicing and operation was often quite inadequate.

A batch of new Fordsons in the USSR in a 1926 picture from the Berghoff album.

Girls from an orphanage near Rostov on Don are given tractor driving instruction in this Berghoff picture.

New tractors were sometimes uncrated and driven away to the farms where they were to be used, without the correct pre-delivery lubrication. The result was that the tractors were sometimes ruined on their way to the farm. There were also reports of drivers filling the fuel tank with crude oil when they were short of petrol, which could result in a damaged engine.

The Krasny Putilowitz tractor factory near Leningrad had been opened in June 1923, but was not fully operational until the following year. It had been equipped with machines which were up to date and mainly of American and British origin.

When the Ford delegation visited the factory in 1926, they reported that the labour force of 800 was producing an average of two tractors a day. The final inspector at the plant, who presumably had a fairly easy job to do, was a former Ford employee who had decided to move from Dearborn, and who now wanted to return to the United States. The Putilow copy of the Fordson had been based on an early version of the American tractor, and was still being built without brakes and with a Holley type vaporiser.

During their tour of farms, members of the Ford delegation were told that the Putilow tractors had a poor reputation for reliability, and the oil consumption was excessive.

Another visitor to the Krasny Putilowitz factory was Peter MacGregor, a senior Ford employee who arrived in 1929 at the request of the Soviet authorities. His report to the factory management was highly critical, as the surviving copy in the Ford archives shows.

Mr MacGregor said that his own 'boss' would close down the factory immediately, so that employees at every level could clean the place up. A tractor and trailer would be needed to haul away the vast amount of junk accumulated over a period of years.

The MacGregor report also expressed considerable concern about the state of morale in the work force. Pay rates should be increased, he advised, with a switch away from payment on a piece-work basis.

When production hold-ups left the machine shop men without work—which appears to have been a fairly frequent occurence—it was recommended they should be sent outside into the yards to work with shovels, instead of being left to stand around idly chatting and smoking. The yard work, with shovels, was left for women to deal with, while the men worked inside.

'I have seen more girls doing more hard manual labour in one hour than I ever saw in all my life. The greatest surprise to me is that the law and the doctors allow it', said the report.

The tractor deals were only a small part of the total business between Ford and the Russians during the 1920s and early 1930s. Trucks and cars were also exported in large numbers, and there was an additional agreement through which Ford planned and equipped two large truck factories for the Russians.

Another Berghoff picture of Fordsons in the USSR.

Henry Ford took a close personal interest in the USSR business, and he appears to have been generous in the terms he agreed with Amtorg. Ernest Liebold, a senior executive with Ford during the period of the USSR deals, stated that Henry Ford was willing to extend credit for the first big tractor purchase, at a time when exporters were being firmly advised to demand immediate payment from the Russians.

Liebold also said that Henry Ford was anxious that his company should play a part in the development of the economy of the USSR. If the working conditions and the standard of living of the ordinary Russians could be improved, he argued, it was less likely that they would want to go to war again. Henry Ford was a committed pacifist.

Nevins and Hill in *Ford: Expansion and Challenge*, quote figures to suggest that the contract agreed with Amtorg for the two truck factories left Ford with a final loss of almost $600,000. 'To give his ideas a practical illustration on the world state, Henry Ford would gladly have sacrificed twice that sum', they said.

In the USSR, Ford cars, trucks and tractors had become a familiar sight in most areas. In 1927 the company was able to claim that 85 per cent of the trucks and tractors there were built by Ford. In the immensely important Ukraine region, there were 5,700 Government owned tractors in 1926, and 5,250 of these were Fordsons, according to a 1927 report.

While Ford products were achieving so much prominence on the fields and roads of the USSR, Henry Ford's industrial methods and achievements were attracting considerable interest and respect among those who were trying to plan the future industrial development of the USSR.

It gave Henry Ford, perhaps the most successful capitalist of his day, a great deal of satisfaction to know that his products and ideas were helping to provide a better standard of living for the ordinary people in the USSR.

Fordson
Fordson

Fordson Evolution

For the Ford organisation, 1926–7 was a crucial period. The Model T car, on which the company's prosperity had been based for nearly 20 years, had become dated. Rival motor manufacturers in the U.S.A. and in Europe had been steadily improving their products and their production methods. The Model T and its truck derivative, the Model TT, were facing increasingly effective competition in both design and price.

The time had come for the fabulous 'Tin Lizzie' to leave the centre of the stage, and towards the end of 1926 Henry Ford was working on the replacement model. May 1927 brought the end of the Model T production run. For a few months thereafter the company had effectively ceased to be a car manufacturer while the inventory of Model Ts was still being sold; and preparations were being made at the Rouge plant in Dearborn to begin manufacturing the new Ford car.

Production started again in December 1927 and the new Ford, the Model A, was announced to the public in an international burst of publicity. The company estimated that 10.5 million Americans saw the car on the launch day, and in London the police were called in to control the crowds who wanted to inspect the new Ford. Souvenir hunters among the Australian first-day crowds almost wrecked the Model A displayed in Melbourne Town Hall.

Amid the various pressures of switching from the Model T car to the Model A, the Fordson tractor was left on the sidelines.

Tractor sales in the U.S.A. were falling at the end of 1927, while the competition from other manufacturers had become more intense. Meanwhile, the Rouge plant was being completely reorganised to accommodate the assembly lines for the new car.

In January 1928, the decision had been taken to stop building the Fordson. The explanation given by the Company was the need to make extra space available for the Model A car, but the world-wide slump in tractor sales probably contributed to the decision.

There is no indication that Henry Ford had any intention of a permanent move out of the tractor business. There is evidence that he was already planning to develop a new tractor to replace the Fordson, and the enthusiastic reception of the Model A car probably encouraged the idea of a replacement tractor. In their book *American Business Abroad*, Mira Wilkins and Frank Ernest Hill state that Henry Ford during a meeting in London with a representative of the Irish Government, referred to the possibility of building a new tractor in the Cork factory.

Although the new tractor design failed to materialise at that time, Henry Ford persisted with the idea and produced a series of experimental designs during the 1930s.

Meanwhile, there was still a substantial market available for the Fordson, and further proof of this fact arrived after production had finished at Dearborn, with another large order from Russia.

New tractors parked near the Cork factory in 1929.

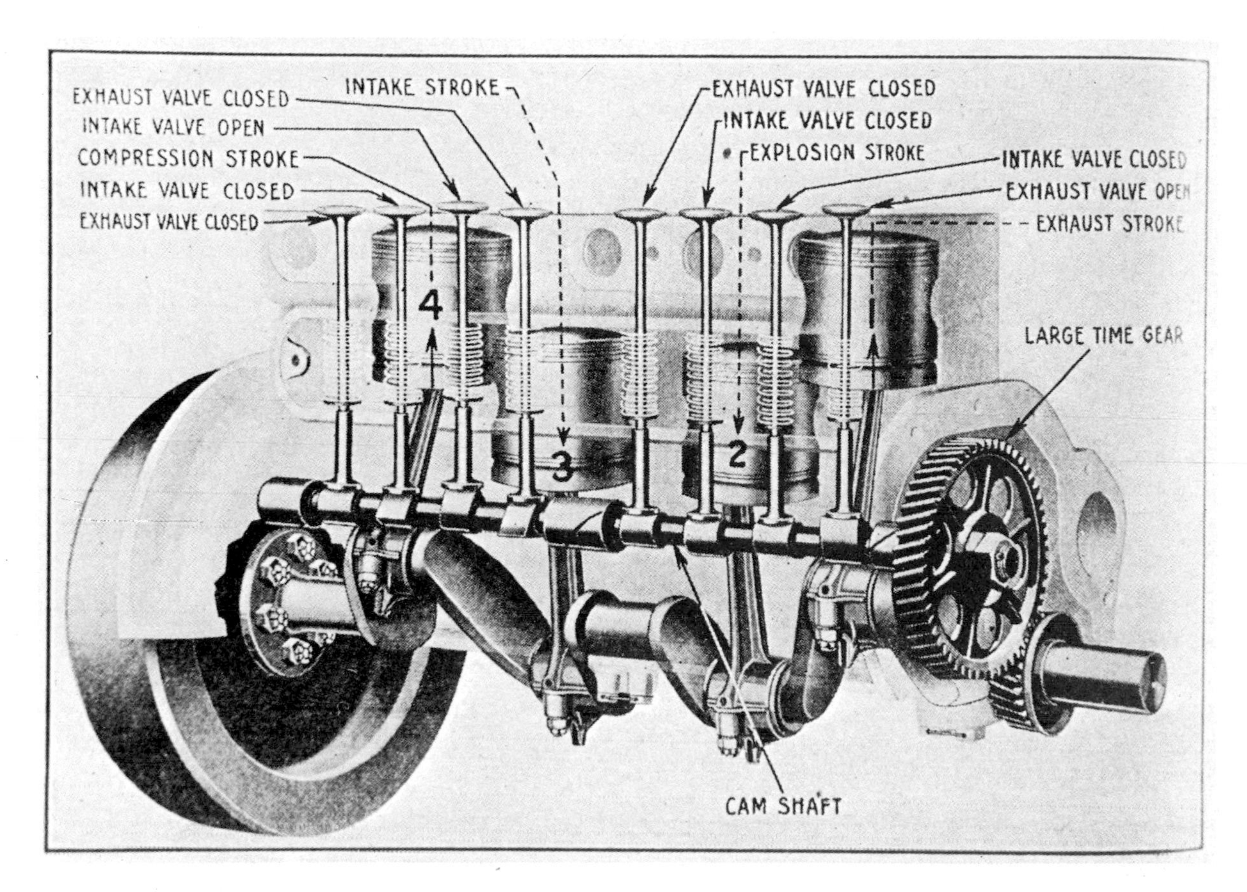

A diagram of the Fordson engine valve arrangement.

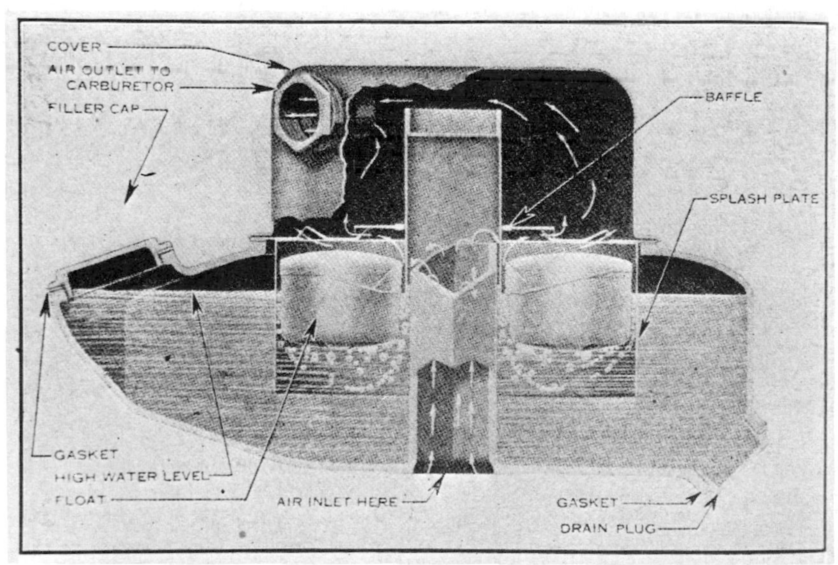

A sectional diagram of the water washer which survived until an oil bath air cleaner replaced it in 1937.

With no tractors available from Ford, the Amtorg purchasing agency turned instead to International Harvester, who were able to meet the order.

The inability to supply the tractors required by the Russians may have encouraged the company to speed up arrangements to resume production. There was also an increasingly urgent need to build up production again in order to ensure a continuing supply of spare parts for the owners of almost 750,000 Model F tractors scattered around the world.

When tractor production was restarted, it was with an improved version of the old Fordson, and at the Cork factory. The decision to centre the whole of Ford's tractor manufacturing operation in Cork was taken during 1928, only six years after the previous attempt to build up large-scale tractor production there had ended in failure. Sir Percival Perry was keen to bring the Fordson back to Cork because there was still a worthwhile market for the tractor in Britain and also in some other countries for which he had the responsibility for Ford business. He was supported by Henry Ford, who still wished to build up his manufacturing business in Ireland, and was also keen to get his company back into the tractor market again.

Equipment from Dearborn was shipped to Ireland during the winter of 1928–9, and production started early in 1929 to meet some urgent orders for spare parts. According to one report there was so much pressure to get production underway that some of the workmen were operating equipment outside in the rain, before it had been moved to its position inside the factory.

A later engine with an oil bath air cleaner.

Side view of a Fordson water washer air cleaner.

Prospects for the new tractor venture appeared to be good. There was a big and continuing demand for replacement parts, and when tractor production started later in 1929 there was a large backlog of orders to satisfy. There was also the prospect of further big orders from the USSR.

Although the Russians had turned to International Harvester in 1928, when they had been unable to buy the Fordsons they wanted, they came back to Ford with further orders after the Cork operation had restarted. In fact, these additional orders were never confirmed because the company was unwilling to accept the credit terms requested by the Russians.

While the new model to which Henry Ford had referred was still several years away from the production line, the move from Dearborn to Cork provided an opportunity to make some important changes to the Fordson.

The 1929 production version was known as the Model N, which began at Cork with serial number 757369 and ended there in 1932 with number 779135. Because the number produced in Ireland was so small, these tractors are of special interest to collectors.

Most of the design changes introduced in 1929 involved the Fordson engine. The cylinder bore was increased by $\frac{1}{8}$ in to raise the power to 27.3 hp according to the British RAC system. The Nebraska test ratings were 23.24 hp at the belt with the kerosene engine, and 29.09 hp on gasoline. The output with the previous 4 × 5 in cylinders of the Model F had been 20.19 hp.

Fordson N final drive arrangement including the worm and worm wheel and the differential.

Opposite below

The familiar Fordson worm and worm wheel.

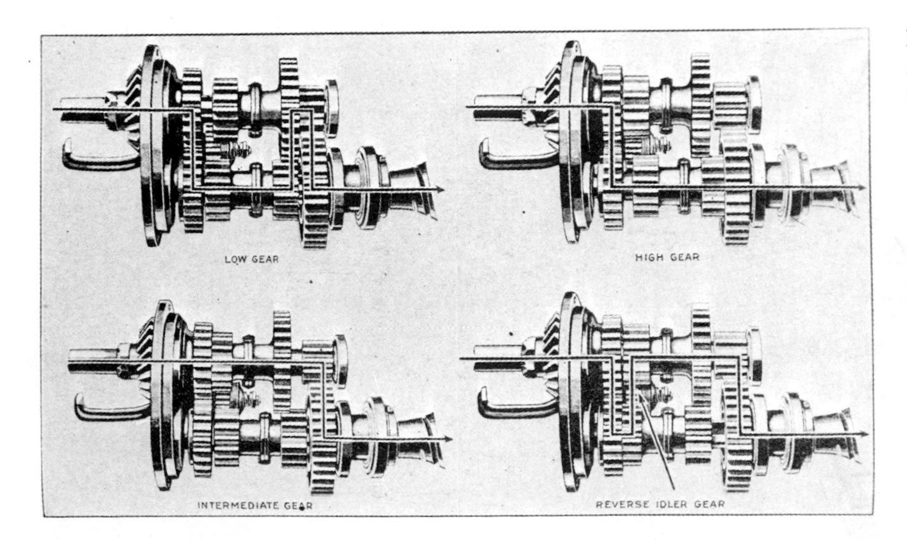

Diagram of the forward and reverse gears in a Fordson N gearbox.

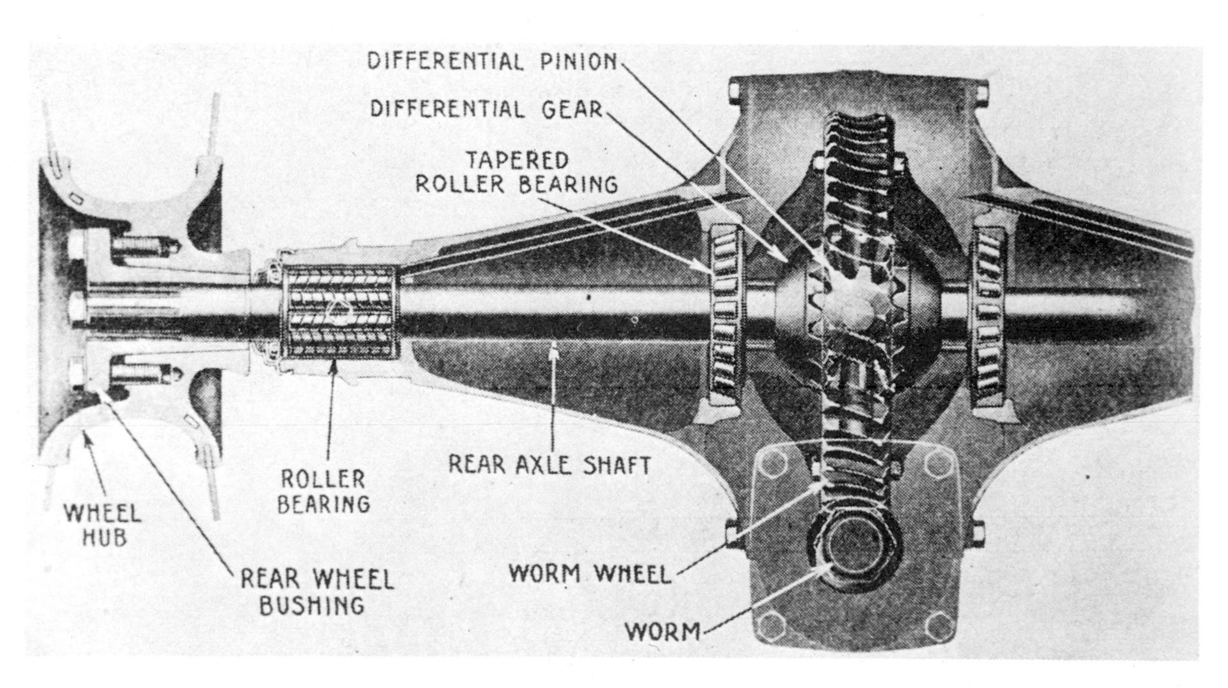
DIFFERENTIAL PINION
DIFFERENTIAL GEAR
TAPERED
ROLLER BEARING
ROLLER
BEARING
REAR AXLE SHAFT
WHEEL
HUB
REAR WHEEL
BUSHING
WORM WHEEL
WORM

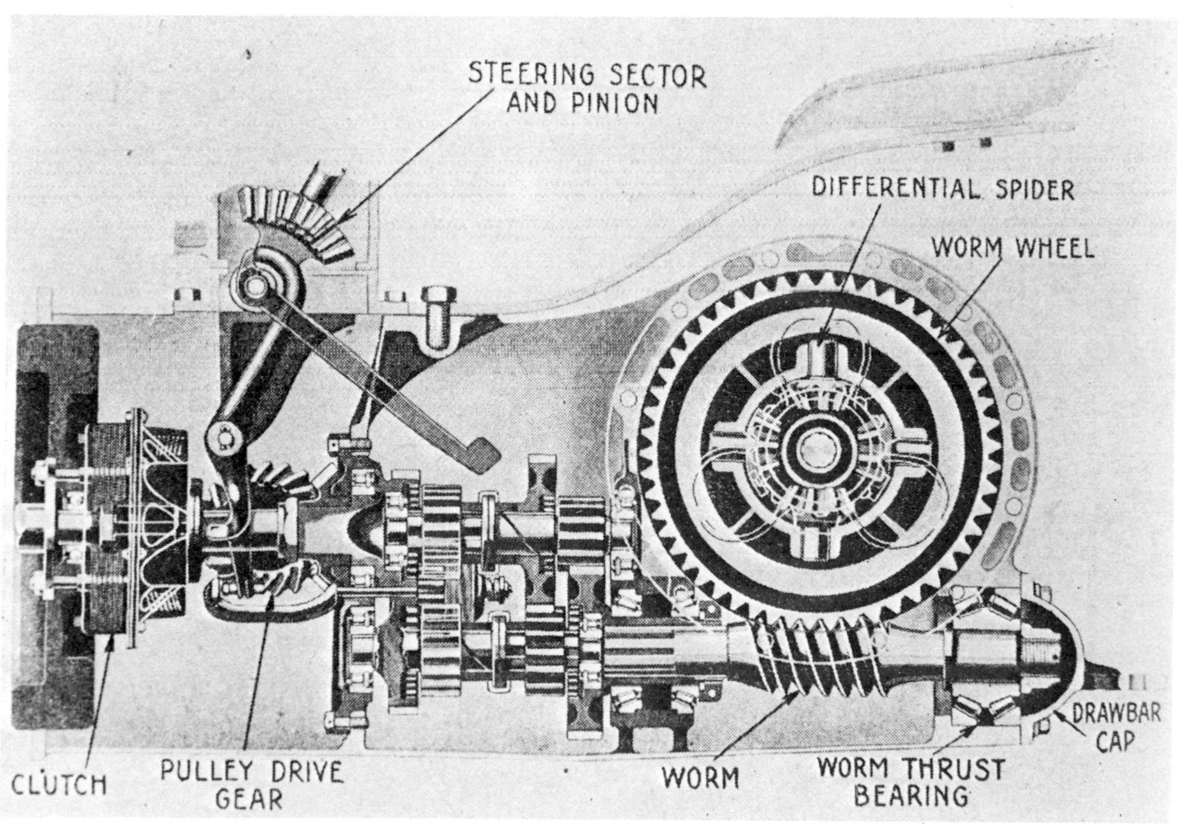
STEERING SECTOR
AND PINION
DIFFERENTIAL SPIDER
WORM WHEEL
DRAWBAR
CAP
CLUTCH
PULLEY DRIVE
GEAR
WORM
WORM THRUST
BEARING

A 1933 photograph of an Irish Model N used in the first public demonstration of the new Dunlop tractor tyres.

External changes included heavier, stronger front wheels with a pattern of five, rounded, triangular holes; the words 'Ford Motor Company Ltd, England, Made in Irish Free State' or 'Made by Ford Motor Company Ltd, Cork, Ireland' on the fuel tank end; and a front axle with a slight bend in the middle. The extended rear wings with a built-in toolbox, which had been an option on Model F tractors from 1924, became standard equipment on tractors from the Cork plant.

The tractor priorities at the Cork factory changed rapidly. During 1929 the biggest problem was trying to catch up with the backlog of orders. Within two years the farm equipment market had slumped and the main concern at Cork was finding sufficient orders to keep the tractor line moving.

Meanwhile the Ford company's interests in Britain were increasingly concentrated in the factory complex at Dagenham, Essex. This was a site which had been chosen in 1923 for a major factory development project, with construction work starting on the marshy land in the 1929 ceremony when Edsel Ford cut the first turf with a silver spade.

Edsel Ford (*left*) with Lord Perry and Henry Ford II during site preparations for the Dagenham factory.

A Model N Fordson with full track conversion.

During 1932, with tractor production running at an all-time Ford low of 50 to 60 a week, Sir Percival Perry again intervened, this time to recommend taking the Fordson out of Cork and into the Dagenham factory. The economic arguments for the move were overwhelming, and with approval from Dearborn, Perry began shipping the factory equipment from Cork to Dagenham during the second half of 1932, and was ready to begin production in the new plant on February 19th, 1933. This left the Cork factory once more as a relatively small assembly unit for cars, operating as a satellite to Dagenham.

While the production transfer was going ahead, it provided an opportunity to give the Fordson another facelift. This time most of the changes showed, including a new dark blue paint finish to replace the previous grey colour; a ribbed pattern on the front of the radiator top tank, with the Fordson name cast into the side panels. The toolbox was moved out of the mudguards and on to the dash panel of the agricultural model. The lettering on the oval fuel tank end plate now read 'Ford Motor Company Ltd. England, Made in England'.

Dagenham production started slowly at first, with sales restricted by the agricultural depression which had caused such problems at Cork. In fact, production in 1934, Dagenham's first full year as a tractor plant, was only 3,582 tractors, little better than the 3,501 production total for 1931, Cork's last full year as a tractor plant.

In 1935, the tractor market started to recover, particularly in the U.S.A., and this was reflected in the Fordson production figures of 9,141 in 1935, which was more than doubled to 18,698 in 1937.

The basic design of the Fordson was 20 years old in 1937, and in spite of the various improvements and the change to blue paint, it was beginning to appear seriously outdated. This was particularly evident in the U.S.A., easily the most important tractor market, where the competition had squeezed the tractor from Dagenham down to less than five per cent of sales. A rowcrop model with tricycle type wheel arrangement to suit the American market arrived belatedly in 1937, known as the Fordson All-Around, and an improved power take-off arrangement from a location on the right hand side of the transmission housing in 1936, were developments which did little to improve the market share figures.

Industrial tractors working at Ford's Manchester plant.

Another attempt to put new life into the old Fordson design came in 1937 with a further round of modifications. One of the most important of these was pensioning off the old water-washer air cleaner to make way for a new oilbath unit. This was certainly a change for the better, as the water-washer was vulnerable to frost damage, and some dust particles were able to get through the system by way of the air bubbles.

This time the paint colour was changed from dark blue to orange, which helped to brighten the landscape, and was probably prompted by American demands for a more conspicuous appearance.

One more important development in the competition for sales was an attempt to raise the power output from the Fordson engine. This had already been achieved once, as part of the 1929 modifications, by increasing the cylinder bore. Another bore increase would have involved a major redesign of the cylinder block, and the sales volume did not justify a big investment. Extra power might also have been gained by speeding up the engine, but there was a risk that the extra rpm could overtax the splash lubrication system. A pressurised lubrication system had been considered at an earlier stage in the engine's development, but had been ruled out then because of the tooling cost.

Ribbed radiator top tank introduced in 1932.

Publicity pictures taken for Ford in Essex during the 1930s.

This left the Ford engineers with a higher compression ratio as the only practical way in which to achieve the extra horsepower the salesmen wanted. It also suited the new Tractor Vapourising Oil (TVO) fuel, available from Shell, with a higher octane rating than that of the standard kerosene used in most 1930s tractors.

The increased compression ratio brought the required power boost, but it also caused a series of engine problems. These included crankshaft, big end and bearing failures, plus excessive fuel and oil consumption. The problems were caused by the wide variations in the quality and octane levels of the various fuels used in the tractors. Not everyone was using TVO, and other fuel grades brought pre-ignition when the high compression ratio cylinder head was fitted. In order to cure the problem, the cylinder head and piston tops were redesigned in 1939, with a special vapourising oil head fitted as standard.

In spite of this short-lived, and quite exceptional, mechanical setback, the Fordson continued to dominate the market in Britain and to provide the Dagenham factory with substantial export business as well. Although the basic design was now well over 20 years old, it was still providing low-cost power which the market required.

Fordson tractors had arrived in Britain in 1917 because war had brought a national emergency. In 1939 it was becoming increasingly obvious that another war on a global scale was a serious possibility. Percival Perry was one of those who had realised the danger at an early stage, and he also remembered how important it had been in the previous emergency to use tractors to help grow more food.

The 150,000th British built Fordson at the end of the Dagenham production line and ready for painting in 1943.

Early in 1939 Perry and a senior executive of the Ford organisation in Britain, Patrick Hennessy, had arranged a meeting with officials from the Ministry of Agriculture and Fisheries. During the meeting they made an extraordinarily far-sighted and, apparently, generous offer to help make additional tractor power available if war was declared.

The proposal was to increase tractor production at Dagenham in order to build up a stock of 3,000 Fordsons. These would be purchased by the British Government at list price less 27.5 per cent, and they would be distributed to Ford dealers' premises in different parts of the country where they would be held in reserve. If war was declared, the Government would have this stock of tractors available for immediate use. If there was no war, the company guaranteed to buy back the tractors.

There was an undertaking in the agreement that the tractors would be kept in working order, and the dealers involved would start the engines from time to time to make sure that the Fordsons in their care were ready for action.

In the worsening international situation, the British Government welcomed the Ford initiative as a positive contribution to increasing Britain's farming efficiency in the event of war. How-

A Fordson towing a load of bombs on a British airfield during the Second World War.

ever, there was a complication in the negotiations when the Ministry of Agriculture and Fisheries heard that a new Ford tractor was under development in the United States.

According to the version of events outlined in *American Business Abroad*, Harry Ferguson had written to the Minister to suggest that the British Government might consider building the new Ford tractor in England. Henry Ford and Harry Ferguson had recently reached an agreement for Ford to develop and build a new tractor incorporating some Ferguson features.

When the Ferguson letter arrived in London, the Ministry there assumed that there might be a plan to replace the Fordson tractor with the new model from the U.S.A. This could mean that the 3,000 Fordsons which the British Government had been offered might very soon become obsolete.

The Minister immediately contacted Perry, who knew little about the American Ford developments, and was probably perplexed by the Minister's questions. Perry was able to give a complete assurance that there were no plans to switch from the Fordson to the American model, which was still under development, and the British Government officially approved the British company's tractor plan on May 18th, 1939.

Roadless tracks on a 1942 green Fordson.

When war was declared in August, most of the 3,000 tractors covered by the agreement were ready for work, and the Dagenham factory was geared up for the big production increase which the war years were to demand.

During the War, with most of the other British tractor companies occupied with military contracts, the Dagenham factory built more than 90 per cent of the agricultural tractors produced in Britain, and by the end of the war well over 80 per cent of the tractors on British farms were Fordsons.

The story of how the Fordson tractor helped to rescue Britain from the threat of starvation during the First World War is well known, and has perhaps been over-stated. Thousands of Fordsons arriving from America to help Britain's farmers save the nation from hunger was an immense psychological boost for war-weary people, and the press gave the new tractors plenty of publicity. But the land the tractors ploughed during the winter and spring of 1917–18 did not yield a harvest until later in 1918, when the end of the War was near.

However, during the Second World War, the situation was quite different. There was less publicity, but the tractors ordered by the Government were actually ready and waiting before the War started, and they were joined by thousands more from the Dagenham factory to make a massive contribution to farming efficiency right through the War.

Rowcrop front wheel arrangement on a 1943 Fordson.

This is an episode in Fordson history which, it seems, has rarely been given the credit it deserves. It is also an outstanding period in the history of the Ford company, as the pre-War rate of tractor production was more than doubled during an extremely difficult period. Dagenham was producing large numbers of engines and vehicles, including Bren Gun Carriers, for the British war effort and the huge factory beside the River Thames was an obvious target for the Luftwaffe. The factory and workers' homes nearby were bombed several times and six people were killed while at work; but tractor production continued.

The 3,000 tractors built for the Government in 1939 were painted orange. Later that year, the colour was changed to green, perhaps because it was a less conspicuous colour and more suitable for wartime, and this paint finish was used until Model N production ended in 1945.

The ploughing-up campaign during the 1939–45 War appears to have been better organised and more successful than its Great War equivalent. Thousands of Land Girls helped out with the tractor driving, and their green jumpers and brown overalls became as familiar in the British countryside as the grey, blue, orange or green Fordson tractors. Ford provided courses on tractor driving and maintenance at the training school at Boreham, Essex, and in spite of the various supply and production problems the company was able to maintain an efficient spares back-up to keep the tractors in action.

An Australian Malcolm Moore loader on an orange Fordson owned by Mal Brinkmann.

The Malcolm Moore loader in raised position on Mal Brinkmann's Fordson.

Rotary Hoes, now Howard Rotavator Co., developed this trench digger around a green Fordson in 1942.

1948 E27N Fordson, minus its air cleaner pipe.

MCT 528

18 Doe Triple-D tractor.

Previous page

17 Fordson Major Diesel. The cab is a more recent addition.

19 Fordson Super Dexta.

20 Ford 5000.

21 The Ford 7000 helped to pioneer turbocharged diesel power in Britain

MARSHALL
County
FC 1174

FORD
401

Opposite
22 The County Forward Control tractor based on a Ford skid unit.

Opposite
23 Ford 8401, designed for the Australian market.

24 The South African 'Lugger', based on a Ford 7610, and built by Armstrong Motors.

25 Ford Versatile 976 articulated tractor.

26 Series 40 was launched at the 1991 Smithfield Show.

Most of the green wartime Fordsons were supplied to farms, but some were bought by the military authorities who required special paint finishes. Some were painted beige for service with the British Army in the North African desert, and others left the factory in undercoat so that they could be painted blue later for service with the Royal Air Force, or camouflaged in green and brown paint for the Army. The RAF also used some Full Track Fordsons, with Roadless tracks, for moving aircraft on Battle of Britain airfields.

While the Dagenham factory was producing the green Model N Fordson, the company's engineers were designing the tractor which might replace it when the War ended. Some of the plans were ambitious and as early as 1944 Sir Patrick Hennessy had a team working on a new type of engine which could be produced in versions to burn gasoline only, gasoline and kerosene, or diesel.

Because of the wartime difficulties of supplies of material, and the need to bring out the new model as soon as possible after the war, the plans were modified. The new model was the E27N Fordson Major, a direct descendant of the old 1917 Fordson with basically the same engine design. The new model was bigger and looked more impressive, and there was yet another paint colour change, this time from green to blue.

An E27N with a Perkins diesel engine pictured at the National Institute of Agricultural Engineering.

The most important mechanical development was getting rid of the old worm and worm wheel final drive which had been such a familiar feature of the old Model F and N Fordsons. This had been given a high priority because the worm wheel was made of costly phosphor bronze, and because it was a drive system which absorbed too much power.

Production started on March 19th, 1945. There were four models, the Standard Agricultural, Land Utility, Row-Crop and Industrial. With a mixed fuel engine the power output was 28.5 hp at 1100 rpm, with a claimed 19.1 hp available at the drawbar. The 1945 prices were £237 for the Standard version on metal wheels, or £285 on rubbers, with £255 for the Row-Crop. The pulley and power take-off were both optional extras.

During a series of tests carried out by the National Institute of Agricultural Engineering, an E27N tractor achieved a continuous work rate of 1.25 acres an hour, ploughing a medium loam stubble with a four-furrow John Deere plough. The furrow depth was 5.1 in, and the fuel consumption was 2.51 gallons per hour, or 2.05 gallons an acre.

County Full Track version of the E27N Fordson Major.

Methane-burning conversion for the Major built for experimental purposes by the National Institute of Agricultural Engineering in about 1949.

Demand for tractor power increased sharply after the War and the E27N sold in large numbers, with production reaching a peak in 1948 with more than 50,000 leaving the Dagenham factory. During its seven-year production span, the tractor achieved an excellent reputation for reliability and durability, and some are still at work.

Like its predecessor, the E27N was used for a long list of attachments and conversions, notably the Roadless half-track kit and the County Full Track or CFT. Hydraulically-operated rear linkage was a later addition, and there was also a factory-fitted diesel version. This used a Perkins P6 diesel engine developing a maximum of 45 hp on the belt at 1,500 rpm—an all-time record for the Farkas based tractor series originating in 1917.

Ford
FERGUSON
SYSTEM

The Handshake Agreement

The decision to close down the Ford tractor operation in the U.S.A. in 1928 had not been planned as a permanent arrangement, and Henry Ford was probably expecting that his company would soon resume its leading position in the American market with a new and more advanced design.

In fact, there was a delay of more than ten years before the new tractor arrived. During this time the company's interest in the world's largest tractor market was limited to a relatively small number of Model N Fordsons shipped from the Cork and Dagenham factories.

One reason for the slow start to the development programme for the new Ford tractor was the depression affecting American agriculture during the early 1930s. Sales of new tractors slumped as tens of thousands of farmers faced financial ruin, and production fell from about 200,000 in 1930 to less than 72,000 in 1931, with a further sharp reduction in 1932.

Ford's motor car production was also suffering through the economic depression, and the output of cars and trucks fell from 1.7 million in 1929 to less than 400,000 in 1932 and 1933. In this situation, finance for a tractor development programme would have been low in the Company's list of priorities.

At some time in the mid-1930s, as the sales graphs were turning upwards again and economic confidence was returning, Henry Ford began experimenting once more with tractor ideas.

A 1937 experimental tractor with a V-8 power unit.

Some of this development work was carried out in a small engineering workshop built in the grounds of Fair Lane, the home Henry Ford had built overlooking the River Rouge in a wooded area he had known well as a child.

The engineer responsible for much of the tractor development work at this stage was Howard Simpson. He had been sent to England in 1932 to make some design changes to the Fordson before production began at Dagenham. From about 1934–35 he was working closely with Henry Ford at Dearborn, and one of the experimental designs he produced brought a particularly enthusiastic response from Ford.

The design which interested Henry Ford so much was a row-crop tractor powered by a V-8 engine, and with front wheels, lights, radiator, hood and steering components taken from the Ford car and truck production lines.

This tractor was completed towards the end of 1937. In the following January it was demonstrated to a group of agricultural journalists. On this occasion Ford was in an optimistic mood and Nevins and Hill in their book *Ford: Decline and Rebirth* report that he was 'as pleased as a small boy with a fire engine'. He is quoted as telling the newsmen, 'I don't care if we can't make a cent of profit. The main thing is to get something started.'

Howard Simpson's V-8 tractor was taken to the Ford farm for evaluation. Meanwhile, Henry Ford's interest had moved on to other ideas and more experimental designs were being developed.

The V-8 was kept for several years as part of the farm tractor fleet. It was later presented to the Henry Ford Museum, where it was restored and maintained as part of the collection of experimental and production tractors.

In 1982, this tractor was included in an auction sale of surplus items and was bought by a private collector. The Museum's display of tractors is a valuable and unique record of the development of Ford tractors since 1907. The V-8 model was an important stage in the progress of Henry Ford's ideas about tractor design, and it is difficult to understand why the Museum Board of Trustees approved the sale of this irreplaceable exhibit.

There is no date for this picture, but the tractor probably dates from Henry Ford's mid-1930s experiments.

The Dearborn tractor experiments were interrupted in October 1938 by the arrival of Harry Ferguson from Britain. Henry Ford had heard about the system of implement attachment and control which Ferguson and his team had developed, and he had suggested that Ferguson might like to bring one of his tractors to the U.S.A. in order to give a demonstration.

Harry Ferguson had made considerable progress since 1918, when he had demonstrated his new mounted plough on the back of an Eros Model T car conversion. His hydraulically-operated, three-point linkage system was being manufactured in Britain on the Ferguson Type A tractor, built under an agreement with David (later Sir David) Brown.

The Ferguson-Brown, as the tractor is usually called, had arrived on the British market in 1936, complete with a range of specially developed implements to operate on the three-point linkage. The performance was good, and the tractor convincingly demonstrated the practical benefits of the Ferguson System.

Henry Ford at the wheel of a 1938 experimental model.

Some of the 1930s tractor development work was carried out in a building at Fair Lane, the Ford Home by the River Rouge.

Harry Ferguson controlled the marketing company, which he organised with great efficiency. There was an excellent service back-up available, and a special training school offered courses for farmers, tractor drivers and dealers in order to help ensure that the new equipment was used correctly.

In spite of the tractor's advantages and the sales efforts, there were not enough customers and the Ferguson marketing organisation was soon facing financial problems.

One reason for the poor sales was the economic difficulties facing the farming industry in Britain and much of Europe in the mid-1930s. Most farmers lacked the finance and the confidence to invest in new equipment. The Ferguson tractor looked smaller and lighter than its principal rival, the British-built Fordson, and many farmers found it hard to believe that such a lightweight machine could, in many situations, out-perform the familiar Fordson.

Price was also a considerable barrier to sales, as the little Ferguson cost £224 while the Fordson sold for about £140. The Fordson used the trailed equipment which many farmers already owned, while the Ferguson required a special set of implements which added substantially to the total cost of the Ferguson System.

A 'Ferguson-Brown', the British built tractor which Harry Ferguson used to demonstrate the Ferguson System to Henry Ford.

The financial problems imposed a strain on the relationship between David Brown and his partner, and this was aggravated when Harry Ferguson refused to accept David Brown's suggestion that design changes were needed in order to make the tractor more saleable. Eventually David Brown decided to go ahead with the modifications he considered necessary, without Ferguson's approval or co-operation.

This was the situation when Harry Ferguson decided to go to Dearborn in 1938. He took a Ferguson-Brown tractor and some implements with him, but he did not tell his partner anything about the purpose of his journey.

Any tractor event in which Harry Ferguson was involved had to be carried out with meticulous care, and the demonstration at Dearborn was no exception. It was held on land near Fair Lane, with Henry Ford as an increasingly interested spectator. Colin Fraser in his book *Harry Ferguson, Inventor and Pioneer* states that

This was a prototype version of the 9N, completed early in 1939.

an Allis-Chalmers B tractor and a Fordson were brought up to the demonstration area from the Ford farm, at Henry Ford's request, so that their ploughing performance could be compared with the Ferguson-Brown.

Henry Ford was impressed by the advantages of the three-point linkage and the extra pulling performance achieved with the system of draft control when the tractors were working up a slope. Harry Ferguson had brought a small, spring-operated model tractor with him and he used this to help explain how the system of implement attachment and control worked.

On Henry Ford's instructions, two chairs and a table were brought into the demonstration field, and he and Ferguson sat in the sunshine to discuss tractors and their own ideas about agricultural mechanisation.

Harry Ferguson (*left*) with Henry Ford and the rear linkage of a 9N.

Demonstrating the Ferguson System's ability to work in confined spaces during the 9N launch.

The two had much in common. Both were farmer's sons, and both had retained a close interest in farming. Henry Ford's ideas about the importance of low-cost, efficient tractor power for the development of the farming industry were very much in line with Harry Ferguson's views.

By the end of the conversation the two had agreed to work together to produce a completely new tractor. The details of the arrangement, involving a massive investment by Ford and the patents which represented 20 years of work by Ferguson and his team, were never witnessed or formally recorded. Instead, the two men simply shook hands on the deal and agreed that they would work together in a spirit of mutual trust.

The handshake agreement, as it was later called, led to the production of a series of immensely successful tractors which made a substantial contribution to the development of power farming throughout the world. It also led to considerable confusion because nobody, perhaps not even the two who had shaken hands, was quite sure what the agreement covered and how it should be interpreted.

Under the terms of the agreement, Ford was given responsibility for manufacturing the new tractor, which would include the new system of implement attachment and control. Harry Ferguson was to be responsible for marketing the tractors built by Ford, together with the implements which were to be sourced from various outside suppliers. It was, apparently, also agreed that the Ford factory in Dagenham would eventually change its production from the Fordson to a new Ferguson System tractor. Another part of the unwritten agreement was an understanding that either side could terminate the agreement at any time and for any reason.

Soon after the demonstration at Dearborn Harry Ferguson returned to Britain to sort out some of his affairs there, in preparation for a long stay in the U.S.A. His involvement with the Ferguson-Brown project was immediately terminated by Ferguson on the grounds that David Brown's decision to introduce design changes had broken the terms of their original agreement.

The Ford and Ferguson names linked at the front of a 9N tractor.

This left David Brown free to develop his own ideas into a new tractor, which was manufactured and sold under his own name. This tractor, the David Brown VAK 1, was launched at the 1939 Royal Show in England. It included the features which David Brown had planned to introduce in a new version of the old Ferguson-Brown design; a more powerful engine, a four-speed gearbox instead of the three ratios Ferguson had insisted upon, and a built-in power take-off. The VAK 1 production schedule was hit by wartime restrictions, but a later version, called the Cropmaster, was so successful that it appears to have vindicated David Brown's views about shortcomings in the Ferguson-Brown design.

Early in 1939, Harry Ferguson returned to the U.S.A. with his family and some of his engineering team. When he arrived at Dearborn he found that development work on the new tractor had already made considerable progress.

Ford 9N engine.

Between the end of October 1938 and the middle of January

Rear linkage of the Ford 9N.

1939, while Ferguson was in Britain, the Ford engineers had stripped down the Ferguson-Brown tractor in order to familiarise themselves with the design of the hydraulic system. They had built three experimental tractors incorporating a three-point linkage system with hydraulic control, and these were being used in a field test programme. Their objective was to design a completely new tractor incorporating Ferguson's ideas, but designed to suit the Ford system of mass production at minimum cost.

The engineering work progressed rapidly in the Ford Rouge factory. Henry Ford was enthusiastic about the tractor project and he made sure that there was no lack of resources to maintain the development schedule.

By the end of March, a further prototype machine, incorporating all the main features of the planned production version, had been completed and was ready for testing. This was taken to Fair Lane for a small celebration party, during which both Ferguson and Ford took the wheel to give a ploughing demonstration for the guests.

There is no doubt that Henry Ford and Harry Ferguson were both determined to maintain the spirit of their handshake agreement, and in general it worked well for many years. But there were disagreements from an early stage in their relationship. This was inevitable when both were so accustomed to being in a position of absolute control.

One example of a difference of opinion in an early stage of the partnership was over the appearance of Ferguson's name on the new tractor. Colin Fraser records Harry Ferguson's own version of events. This claimed that Henry Ford had insisted on the Ferguson name being displayed, and that Ferguson had at first refused and insisted that he had 'no ambitions in that direction'. As Colin Fraser points out, this appears to be quite out of character and it is more likely that Ferguson was anxious to have his name displayed with considerable prominence.

Detail of rear linkage showing the compression spring for the top link.

Charles Sorensen in *Forty Years with Ford* claims that 'Ferguson induced Mr Ford to put his name "Ferguson" on this tractor jointly with Mr Henry Ford's'. Sorensen adds that Edsel Ford and others in the company were against the use of the Ferguson name on the tractor, and that the small 'Ferguson System' plaque which was displayed below the familiar Ford badge, was devised by the Ford patent expert, John Crawford, in order to satisfy Ferguson's 'demands for recognition'.

The tractor was marketed as the Ford 9N, but it was correctly known as the 'Ford tractor with Ferguson System' or, more briefly, as the Ford-Ferguson.

While the engineering work progressed, Ferguson had set up a new marketing company. This was financed partly by a $50,000 loan from Henry Ford, and was called the Ferguson-Sherman Manufacturing Corporation. The Sherman brothers, Eber and George, sold American built Fordson tractors in the 1920s, and had also helped Ferguson to market a plough he had designed for the Fordson. When Fordson production had been transferred to Dagenham, the Sherman brothers had formed a company to import the tractors for distribution in the U.S.A.

Harry Ferguson ended the Ferguson-Sherman Company in 1941, after a disagreement with the Sherman brothers, and took complete personal control of the marketing company which replaced the previous partnership.

In June 1939, there were some production models of the new Ford tractor available, and these were demonstrated to the dealer organisation at an event organised by the Ferguson-Sherman Company. There was another, very much bigger event on June 29th when nearly 500 guests, including overseas visitors and journalists, were invited to lunch and a demonstration on the Ford farm at Dearborn.

An event in the programme which attracted considerable interest was the classic Ferguson demonstration of ploughing all the ground in a 20 × 27 ft enclosure, and leaving no wheelmarks. One of the star performers at the 9N launch was David McLaren, an eight-year old boy from the local Greenfield Village School. He was 'lifted' on to the tractor seat in order to show how easy it was to operate the controls. According to one report, the furrows ploughed by David McLaren were as straight as those which Harry Ferguson had ploughed previously.

Eight-year old David McLaren takes the wheel during the launch programme for the 9N.

Ford 9N tractor at the Henry Ford Museum.

The launch price of the new Ford tractor was $585. This included rubber tires, a power take-off, and an electrical system with a generator and battery, which were all standard equipment on the 9N. During the Second World War, when tractor production was hit by shortages of raw materials, the standard specification had to be reduced by fitting steel wheels and replacing the generator and battery with a magneto. This version was called the Ford 2N.

Optional extras for the 9N included a set of lights, steel wheels with detachable lugs, and a belt pulley. It was also suggested that the tractor's low noise level made it suitable for installing a radio, which was a novel idea in 1939.

Ford's power unit was a four-cylinder, L-head design with 120 cu in displacement. There were three forward gears and a reverse, and the maximum forward speed was 6 mph at the rated 1400 rpm governor setting.

It is difficult to measure the tractor's success in commercial terms because wartime restrictions reduced the production level. This is indicated by the production figures, which fell from 42,910

8N tractor at the Henry Ford Museum.

in 1941 to less than 16,500 in the following year when production was completely stopped for several months. The peak production figure for the 9N was 74,004 in 1946, the tractor's last full year.

For Harry Ferguson the 9N was immensely important. It was the tractor which publicised and proved his ideas on a world scale, and it established him as a person of great wealth and influence in the international tractor industry.

Harry Ferguson had expected the British Ford company to stop manufacturing the Fordson tractor and to replace it with the 9N. He had also hoped to be made a director of the British company.

Henry Ford had firmly supported Ferguson's attempts to influence the British company, and he put considerable pressure on Lord Perry in an effort to persuade him to change his policy towards Ferguson.

In 1939, with war an imminent prospect, it would have been difficult and perhaps impossible to abandon the existing product and introduce a completely new one. This was not the situation after the War when the British company replaced the Model N with the much improved E27N tractor. There is some evidence

Ford

8N tractors on the Ford assembly line.

that the British directors were reluctant to involve Ferguson in their company. He had a reputation for being difficult to deal with and disruptive, and his generally harmonious relationship with Henry Ford was an exception in a business career where broken partnerships seemed frequent.

Meanwhile, in Dearborn there were momentous changes in the Ford organisation. Edsel, Henry Ford's son, died in 1943, and Henry Ford took over as president. Henry Ford had plenty of experience of running the company he owned, but in his 80th year the strains of controlling a huge corporation were beyond his capability. The company's affairs suffered a considerable setback for two years until September 1945 when the frail 82-year old president resigned so that his 28-year old grandson, Henry Ford II, could take his place.

Henry Ford, the man who had done so much to make cheap, efficient tractor power available on the world's farms, died in April 1947 at the age of 83. In the same year, the total production of Fordson and Ford tractors since 1917 reached 1,700,000.

When Henry Ford II took control of the company in 1945, he had to take some tough decisions in order to reverse the declining fortunes of the Ford organisation. One of the problem areas was the American tractor operation which had been losing money, and which was also in the unsatisfactory position of having little control over the marketing of its products.

Various alternatives were considered, some of them involving the sale of a share in the Ferguson distribution company to Ford Motor Co.

At the end of 1946, the Ford tractor plans were announced. An improved version of the 9N would be introduced, and the Ford Company would set up its own marketing organisation, called Dearborn Motors, to handle the new tractor. This left Harry Ferguson with a large marketing company in the U.S.A., but with no tractors to distribute once the Ford company's six months' notice period had expired at the end of June 1947.

The updated Ford tractor, called the 8N, was launched in July 1947. There were more than 20 design improvements, and the most important of these was a new gearbox with four speeds instead of the three available in the 9N. The new model was also equipped with a complete Ferguson System of implement linkage and hydraulics.

One of the results of this action by the Ford Company was the decision by Ferguson to set up his own tractor manufacturing operation in the U.S.A. in order to maintain a place in the American market. In the meantime, he was able to supply enough tractors from his British factory to keep his distribution company operating on a reduced scale.

Another result was the Ferguson lawsuit, with claims amounting eventually to about $340 million for alleged patent infringements and damage caused by the loss of business through the changed tractor marketing arrangements.

The suit was filed in January 1948 and it continued until a settlement was agreed in April 1952, with about one million documents and some 200 lawyers involved in the litigation.

One of the reasons for the complexity of the lawsuit was the original handshake agreement on which the tractor production and marketing arrangements had been based. This agreement was about ten years old by the time the lawyers were trying to determine its implications, and one of the two men who had made the agreement was already dead.

A Ford 8N with a mounted two-furrow plough.

Easily the biggest item in the claim was the sum of $240 million for the alleged damage to the Ferguson marketing organisation resulting from the loss of the Ford tractor business to Dearborn Motors. The balance consisted of claims for patent infringement through the use of the Ferguson System on the new tractor.

When the long, tedious and costly legal proceedings had run their course, Harry Ferguson and his companies accepted a payment of just $9.25 million. This was to cover the use of some Ferguson patents in the 8N tractor. There was also an agreement that design changes would be made so that the new tractor no longer conflicted with the Ferguson patents.

The claim for damages because of the withdrawal of the Ford tractor distribution business was dismissed. This was partly because the Ferguson operations in the U.S.A. were prospering again by the time the case ended, and also because of the Ford claim that they were entitled to terminate the arrangement under the terms of the original handshake agreement.

Of the $9.25 million award, about $4 million was required to cover the Ferguson legal costs, and there was a $500,000 deduction to cover payments to two companies which had also been involved in the case.

In financial terms, the settlement must have seemed a bitter disappointment to Harry Ferguson, but he announced the result as a great victory. He had employed specialists to handle the public relations aspects of the case, and had been able to gain some useful publicity for his tractors.

Meanwhile, the Ford 8N tractor had become a massive commercial success with production exceeding 100,000 a year in 1948 and 1949, matching the best output figures for its predecessor, the Fordson Model F.

Ford 961 POWERMASTER

Powered by Diesel

In 1951, the Ford factory at Dagenham was still building the E27N Major, a direct descendant of the 1917 Fordson, and the 8N, based on the 9N design of 1939, was being built at the Dearborn plant. Although both tractors were still selling in large numbers, with a combined production total of around 140,000 a year, the demands of the market were beginning to change.

One of these changes was the type of engine used to provide power on the farm. Diesel engines were still a rarity, even though they had been available in the tractor market since the 1920s.

In spite of significantly better fuel efficiency available from diesel-powered tractors, sales had remained very low. This was partly because of the additional cost, and it was also because multi-cylinder diesels had acquired a reputation for being difficult to start. An additional factor in the U.S.A. was the low cost of gasoline which had meant that fuel economy was low down in the list of priorities for farmers choosing a new tractor.

During the late 1940s, there were indications that diesel power was beginning to attract increased interest in Britain. The trend was slow at first, but it was given a substantial boost in 1948 when Ford announced that the Perkins P6 diesel was available as an option for the E27N. It was a development which encouraged many farmers to consider a diesel tractor for the first time, and it made the E27N easily Britain's most popular diesel model.

In spite of its commercial success, the diesel-powered E27N had

Simms

Opposite
The New Major ElADKN in a 1952 photograph.

A new Ford emblem for the Fordson New Major series.

Opposite
Ford's new diesel engine for the New Major.

been introduced as a temporary measure. Since 1944 engineers at Dagenham had been developing a completely new engine; by 1951 this was ready for the market.

The 1944 project had been authorised by Patrick Hennessy, and he had continued to support and encourage the development work in spite of some strong opposition from within his own company. The project was controversial because it involved diesel power at a time when there was little evidence that this type of engine would have any substantial commercial future in the tractor market. Allocating scarce engineering resources to a diesel engine development programme was remarkably far-sighted, especially as the engineers were attempting to design a new type of diesel engine.

Ford launched their new engine in an entirely new tractor at the 1951 Smithfield Show in London, and it was in full production from the beginning of 1952. The tractor was the New Major, still sold under the Fordson name, but with new curved styling which helped to set a design trend in the market.

This four-wheel drive tractor from County was built from 1954 on a New Major skid unit.

A New Major diesel was used for experimental work by the National Institute of Agricultural Engineering. This 1965 version has a hydrostatic transmission and automatic guidance.

The Doe Triple-D tractor arrived on the market in 1957.

The New Major feature which attracted most interest was the power unit. It was the engine which Patrick Hennessy had encouraged through its development stages, and it fully justified his optimism. The power unit was available in gasoline, gasoline-kerosene and diesel versions, all using the same basic 3.6 litre block, but with the compression ratio ranging from 4.35 : 1 in the kerosene version to 16 : 1 for operating on diesel fuel.

One result of using the same basic unit for all three engine variants was that a high production volume helped bring down the cost of the diesel. When the production version was announced the company was able to describe it as 'the cheapest diesel of its class in the world'. It was also the first tractor diesel to provide really easy starting, and it was exceptionally reliable and long-lasting.

Ford's new American built 801 and 901 tractor series were announced in November 1957.

The 961 Powermaster with 44 hp at the drawbar, pictured in 1957.

Ford
POWERMASTER

The new Ford engine did a great deal to overcome farmers' objections to diesel power. Towards the end of the production run of the New Major, and of its Super Major and Power Major derivatives, well over 90 per cent of British buyers were specifying the diesel version.

At the Dearborn factory, production of the 8N model ended when the new NAA tractor was introduced in 1953. This was a 30 hp model which was known as the Golden Jubilee tractor to celebrate the Ford Motor Company's 50th year.

Diesel engine options were available on American built Ford tractors from 1958, following the lead given by the Dagenham factory.

A publicity picture of the Fordson Dexta.

The Super Dexta working with a Howard Rotavator.

The Ford 5000.

Besides the increasing importance of diesel power, another significant post-War tractor trend was the need for a range of different models in order to meet the increasingly varied requirements of the market. For almost 40 years, the Ford factories had been successful with a single model policy. In fact, it had been an advantage because the almost total standardisation had been an important factor in keeping costs down and sales up. The policy had been an important factor in the company's commercial success, but by 1955 it was necessary to offer farmers a choice of engine sizes and specifications.

During the 1950s, the Ford tractor range was expanded at both the Dagenham and Dearborn factories. In 1955, American farmers were offered a choice of five different Ford tractor models, with additional low specification and rowcrop versions available. An eye-catching grey and red colour scheme was introduced with the Powermaster series. The colour scheme in Britain had continued to be blue and orange until 1962 when the company standardised on blue and grey.

As the New Major established itself as a big tractor in 1950s terms in Britain, Dagenham added the Dexta series, later uprated as the Super Dexta, in order to cover the highly competitive small tractor market.

Turbocharging arrived on the British tractor market with the Ford 7000.

Another indication of the increasingly varied requirements in the tractor market was the development of specialised versions of Fordson and Ford tractor models. For years, Fordson tractors had provided a mobile power source which other companies used for a wide range of industrial and agricultural equipment, and the Rotary Hoes Trench Digger developed for wartime drainage work is one of many examples, which included shunting units for moving trucks in railway goods yards, scrapers and dozers for site work, and mowers for parks and sports fields.

During the 1950s, there was more emphasis on developing specialised agricultural tractors based on Ford skid units. Four-wheel drive is one example of this, and British interest in the idea achieved its first significance with the County and Roadless tractors produced in collaboration with Ford. County Tractors is still a major producer of four-wheel drives powered by Ford, using equal sized driving wheels and including the unique forward control model.

Both County and Roadless had moved into the four-wheel drive market after specialising in tracklayers, and the County Full Track and Roadless half-track had been available as adaptions of the E27N Major.

The factory at Basildon, Essex builds the Series 40 tractor range for the N.H. Geotech group.

As well as the extra efficiency of tracks and four-wheel drive there was also a demand for additional power. An interesting development to meet this need was the Doe Triple-D tractor, in production at Maldon, Essex from 1957. The Triple-D, short for Doe Dual Drive, was based on a farmer's invention, and consisted of two Fordson New Major tractors joined together to make one articulated machine.

The Triple-D was superseded as other manufacturers brought out new models to meet the demand for extra power. Ford broke through the 100hp barrier in 1968 with the 105hp American built 8000, followed in 1969 by the 9000 powered by a 130hp turbocharged engine.

UK tractor production was moved from Dagenham in 1964 to a new factory 20 miles away at Basildon. The 100-acre site provided space to expand and develop the tractor operations, and the move coincided with the launch of a completely new tractor series, the 2000, 3000, 4000 and 5000 models which provided Ford with a new world-wide range.

The demand for more power brought further developments, and in 1971 the 83hp Ford 7000 became the first tractor on the British market designed with a turbocharger. The TW tractor series took the Ford range up to 186hp, and an agreement to sell Steiger tractors in Ford colours brought tractors of 300hp plus into the range. This arrangement ended when Case IH took over the Steiger company in 1986, and in the following February Ford agreed to buy the

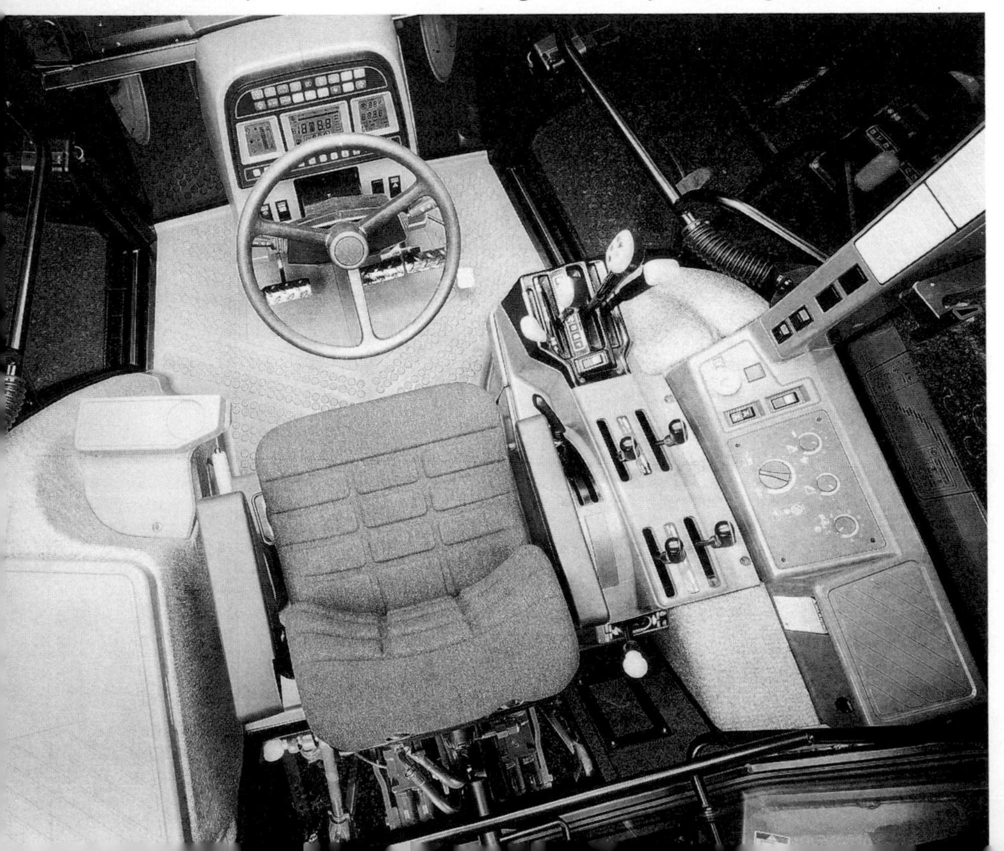

Series 40 SLE cab interior.

Versatile company based at Winnipeg, bringing some of the world's most powerful articulated tractors into the Ford range.

Ford made headlines in October 1985 by taking over the Sperry New Holland machinery business with headquarters in New Holland, Pennsylvania. The deal, which cost Ford about $330 million, added one of the leading ranges of harvesting machinery to the Ford tractor line, and it also brought a change of name to Ford New Holland.

The headlines were even bigger in July 1990 when the Ford and Fiat companies agreed to merge the Ford New Holland and FiatGeotech farm and construction equipment divisions under the N.H. Geotech name, with Fiat holding a dominant 80 per cent interest. The result is the western world's biggest tractor manufacturer, with headquarters in London and with the Basildon factory scheduled to be the group's leading tractor and diesel engine manufacturing facility.

While negotiations between Ford and Fiat were in progress, Ford New Holland engineers were working on a completely new tractor range to cover the top selling 75 to 120hp sector of the market. The Series 40 range was launched at Smithfield Show in December 1990, following a £250 million research and development programme.

There were six models in the first phase of the Series 40 programme, and apart from the Ford badge and the familiar blue paint colour virtually everything on the tractors was completely new.

The development project included a new range of four and six cylinder engines known as the PowerStar series. They offer outstanding power and torque characteristics, and they have been designed with up to 25 per cent more cubic capacity than some of their competitors. A bigger engine means the rated power output is developed at lower speeds, and reduced stress and wear encourage improved reliability and a longer engine life. The design also helps to reduce noise and vibration, and achieves cleaner exhaust emissions and improved fuel efficiency.

Gearbox options for Series 40 tractors include the new ElectroShift transmission providing 16 speeds forwards and in reverse, with push-button shifting, full synchromesh and the choice of 30 or 40kph top gear speed. There is a new hydraulic system, new electronics and instrumentation, and the SuperLux cab on SL and SLE specification models is also completely new, providing Series 40 tractor drivers with a quieter, more efficient work environment for the 1990s.

Appendix 1
How the Fordson Changed

The following photographs show how the Fordson developed through more than 30 years, from the Model F to the E27N.

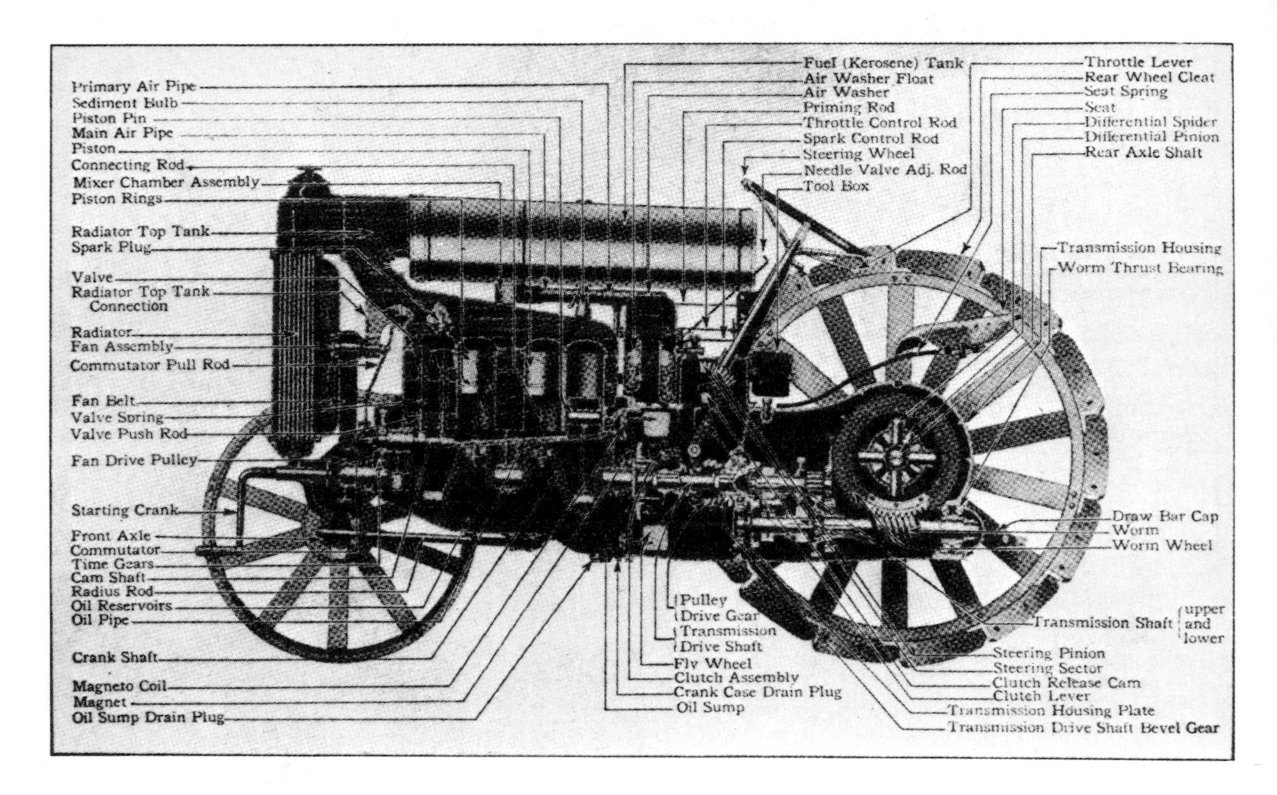

Model F in the early 1920s.

Water washer version of the Model N in 1932.

Opposite top
1938 version of the Model N.

Opposite
The final stage of Fordson development in the 1947 E27N tractor.

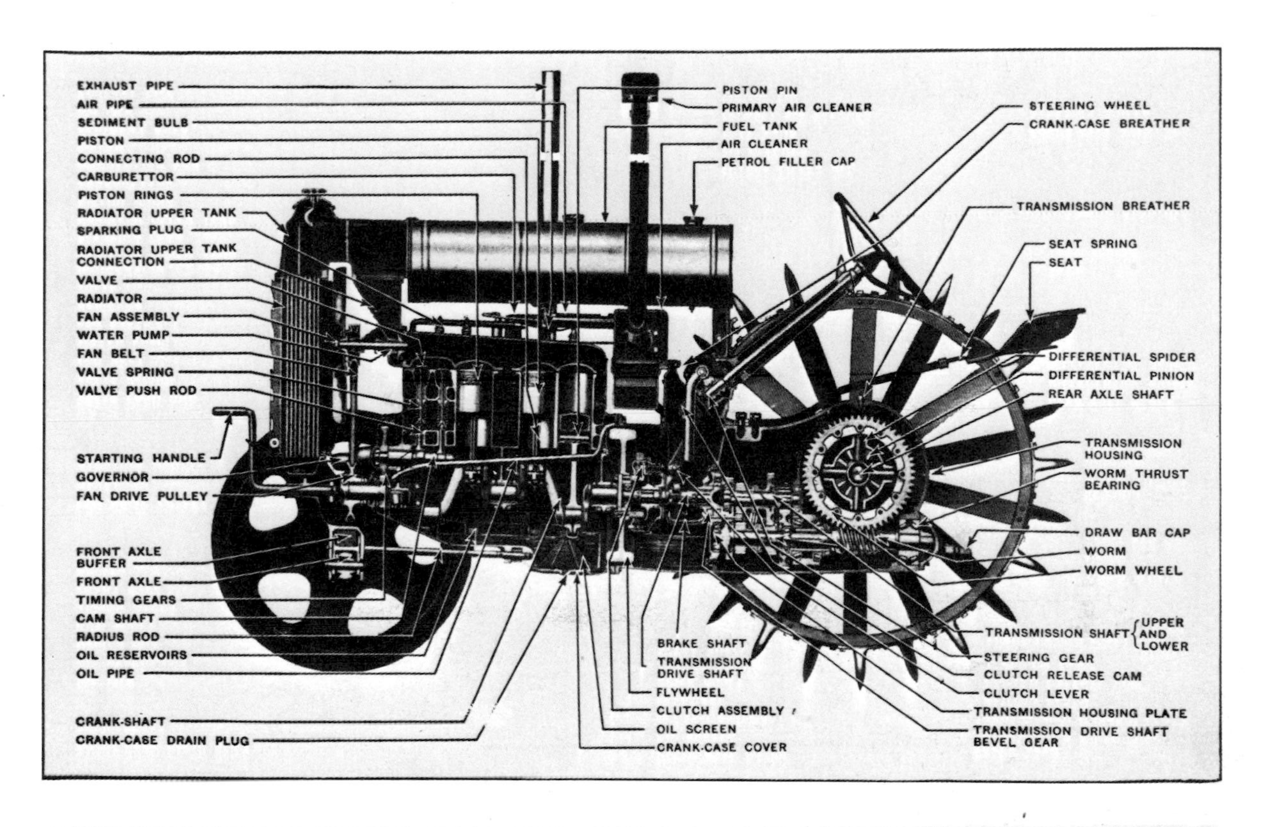

EXHAUST PIPE
AIR PIPE
SEDIMENT BULB
PISTON
CONNECTING ROD
CARBURETTOR
PISTON RINGS
RADIATOR UPPER TANK
SPARKING PLUG
RADIATOR UPPER TANK CONNECTION
VALVE
RADIATOR
FAN ASSEMBLY
WATER PUMP
FAN BELT
VALVE SPRING
VALVE PUSH ROD
STARTING HANDLE
GOVERNOR
FAN DRIVE PULLEY
FRONT AXLE BUFFER
FRONT AXLE
TIMING GEARS
CAM SHAFT
RADIUS ROD
OIL RESERVOIRS
OIL PIPE
CRANK-SHAFT
CRANK-CASE DRAIN PLUG
PISTON PIN
PRIMARY AIR CLEANER
FUEL TANK
AIR CLEANER
PETROL FILLER CAP
BRAKE SHAFT
TRANSMISSION DRIVE SHAFT
FLYWHEEL
CLUTCH ASSEMBLY
OIL SCREEN
CRANK-CASE COVER
STEERING WHEEL
CRANK-CASE BREATHER
TRANSMISSION BREATHER
SEAT SPRING
SEAT
DIFFERENTIAL SPIDER
DIFFERENTIAL PINION
REAR AXLE SHAFT
TRANSMISSION HOUSING
WORM THRUST BEARING
DRAW BAR CAP
WORM
WORM WHEEL
TRANSMISSION SHAFT {UPPER AND LOWER
STEERING GEAR
CLUTCH RELEASE CAM
CLUTCH LEVER
TRANSMISSION HOUSING PLATE
TRANSMISSION DRIVE SHAFT BEVEL GEAR

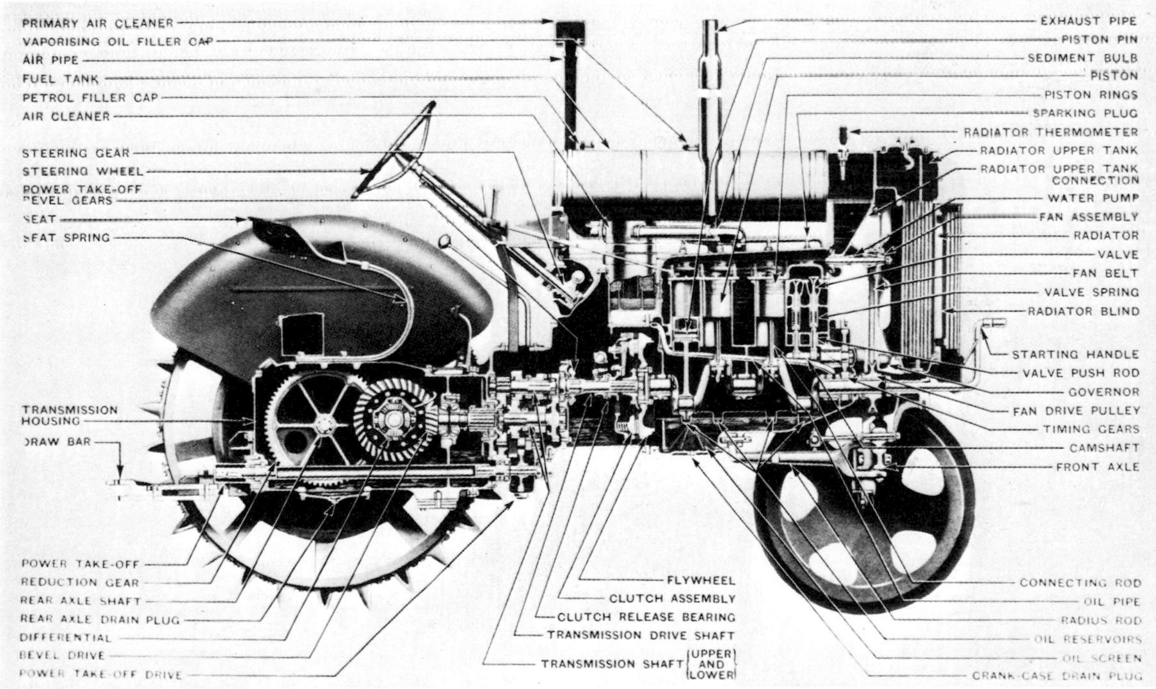

PRIMARY AIR CLEANER
VAPORISING OIL FILLER CAP
AIR PIPE
FUEL TANK
PETROL FILLER CAP
AIR CLEANER
STEERING GEAR
STEERING WHEEL
POWER TAKE-OFF BEVEL GEARS
SEAT
SEAT SPRING
TRANSMISSION HOUSING
DRAW BAR
POWER TAKE-OFF
REDUCTION GEAR
REAR AXLE SHAFT
REAR AXLE DRAIN PLUG
DIFFERENTIAL
BEVEL DRIVE
POWER TAKE-OFF DRIVE
FLYWHEEL
CLUTCH ASSEMBLY
CLUTCH RELEASE BEARING
TRANSMISSION DRIVE SHAFT
TRANSMISSION SHAFT {UPPER AND LOWER}
EXHAUST PIPE
PISTON PIN
SEDIMENT BULB
PISTON
PISTON RINGS
SPARKING PLUG
RADIATOR THERMOMETER
RADIATOR UPPER TANK
RADIATOR UPPER TANK CONNECTION
WATER PUMP
FAN ASSEMBLY
RADIATOR
VALVE
FAN BELT
VALVE SPRING
RADIATOR BLIND
STARTING HANDLE
VALVE PUSH ROD
GOVERNOR
FAN DRIVE PULLEY
TIMING GEARS
CAMSHAFT
FRONT AXLE
CONNECTING ROD
OIL PIPE
RADIUS ROD
OIL RESERVOIRS
OIL SCREEN
CRANK-CASE DRAIN PLUG

Appendix 2
Production Data

The following table shows the details of Fordson and Ford tractor production for the first forty years.

Year	*U.S.A.*	*Cork*	*Dagenham*	*Total*
1917	254			254
1918	34,167			34,167
1919	56,987	303		57,290
1920	67,329	3,626		70,955
1921	35,338	1,433		36,781
1922	66,752	2,233		68,985
1923	101,898			101,898
1924	83,010			83,010
1925	104,168			104,168
1926	88,101			88,101
1927	93,972			93,972
1928	8,001			8,001
1929		9,686		9,686
1930		15,196		15,196
1931		3,501		3,501
1932		3,088		3,088
1933			2,778	2,778
1934			3,582	3,582
1935			9,141	9,141
1936			12,675	12,675

Year	*U.S.A.*	*Cork*	*Dagenham*	*TotaL*
1937			18,698	18,698
1938			10,647	10,647
1939	10,233		15,712	25,945
1940	35,742		20,276	56,018
1941	42,910		22,210	65,120
1942	16,487		27,650	44,137
1943	21,163		26,300	47,463
1944	43,444		23,845	67,289
1945	28,729		17,770	46,499
1946	74,004		25,290	99,294
1947	85,589		34,915	120,504
1948	103,462		50,561	154,023
1949	104,267		33,375	137,642
1950	97,956		42,275	140,231
1951	98,442		35,868	134,310
1952	82,041		30,444	112,485
1953	72,543		29,575	102,118
1954	51,490		45,689	97,179
1955	66,656		48,872	115,528
1956	39,056		40,991	80,047
1957	39,685		46,114	85,799
Totals	1,853,876	39,076	675,253	2,568,205

The following comparison, for 1984, is of interest.

Country	*Total*
USA (Romeo)	19,766
Mexico	4,379
Brazil	7,925
India	8,611
U.K. (Basildon)	25,695
Belgium (Antwerp)	1,664
World	68,040

Below is a listing of the Fordson serial numbers 1917 to 1952, i.e. to the end of E27N production.

Year	*Dearborn*	*Cork*	*Dagenham*
1917	1 to 259		
1918	260 to 29979		
1919	34427 to 88088	63001 to 63200	
		65001 to 65103	
1920	100001 to 158178	65104 to 65500	
		105001 to 108229	
1921	153812 to 170891	108230 to 109672	
1922	201026 to 262825	109673 to 110000	
		170958 to 172000	
		250001 to 250300	
		253001 to 253552	
1923	268583 to 365191		
1924	370351 to 448201		
1925	455360 to 549901		
1926	557608 to 629030		
1927 } 1928 }	629830 to 747681		
1929		747682 to 757368	
1930		757369 to 772564	
1931		772565 to 776065	
1932		776066 to 779135	
1933			779154 to 781966
1934			781967 to 785547
1935			785548 to 794702
1936			794703 to 807580
1937			807581 to 826778
1938			826779 to 837825
1939			837826 to 854237
1940			854238 to 874913
1941			874914 to 897623
1942			897624 to 925273
1943			925274 to 957573
1944			957574 to 975418
1945			975419 to 993488
1946			993489 to 1018978
1947			1018979 to 1054093
1948			1054094 to 1104656

Year	Dearborn	Cork	Dagenham
1949			1104657 to 1138032
1950			1138033 to 1180609
1951			1180610 to 1216574
1952			1216575 to 1216990

Index

Figures in italic refer to black and white photographs